Enhancing Modern Living with Virtual and Augmented Reality

This book explores the role of virtual and augmented reality (VR and AR) in today's evolving digital landscape, examining how these technologies are shaping society and how they might evolve in the future.

It surveys the wide-ranging applications of VR and AR across industries, showcasing their functionality in areas such as entertainment, culture, retail, sports, tourism, and well-being. Through practical examples, the book highlights the benefits of these technologies while also addressing their limitations, challenges, and risks. Critical discussions include the technical, ethical, and societal implications of AR and VR, with a focus on pressing issues such as data privacy, cybersecurity, accessibility, the digital divide, and the psychological effects of immersive experiences. This balanced approach provides a comprehensive understanding of both the opportunities and pitfalls of these technologies.

This volume is an invaluable resource for students, researchers, and professionals interested in the interdisciplinary nature of AR and VR. It bridges fields such as computer science, psychology, media studies, tourism, and urban studies, offering insights into how these technologies are reshaping industries and everyday life.

Kinga Stecuła, PhD, works as a Lecturer in the Faculty of Organization and Management within the Department of Production Engineering at Silesian University of Technology, Poland.

Radosław Wolniak is a full professor in the Department of Economics and Informatics at Silesian University of Technology, Poland.

CRC Press Reference Books in Computer Science

CRC Press Reference Books in Computer Science series offers advanced and authoritative resources for students, researchers, and professionals across the field of computer science. Books in this series serve as indispensable references which present rigorously researched and innovative approaches to current and complex issues.

Reimagining Literacy in the Age of AI: Theory and Practice
Edited by Jason D. DeHart, Suriati Abas, Raúl Alberto Mora, and Damiana Gibbons Pyles

Generative AI in Higher Education: Guiding Principles for Teaching and Learning: Volume 1
Emmanuel K Nartey

Governing the Future: Digitalization, Artificial Intelligence, Dataism
Henning Glaser and Pindar Wong

Enhancing Modern Living with Virtual and Augmented Reality
Practical Applications
Kinga Stecuła and Radosław Wolniak

Enhancing Modern Living with Virtual and Augmented Reality

Practical Applications

Kinga Stecuła and Radosław Wolniak

CRC Press
Taylor & Francis Group
Boca Raton London New York

CRC Press is an imprint of the
Taylor & Francis Group, an **informa** business

Designed cover image: Shutterstock

First edition published 2026
by CRC Press
2385 NW Executive Center Drive, Suite 320, Boca Raton FL 33431

and by CRC Press
4 Park Square, Milton Park, Abingdon, Oxon, OX14 4RN

CRC Press is an imprint of Taylor & Francis Group, LLC

Library of Congress Cataloging-in-Publication Data
Names: Stecuła, Kinga author | Wolniak, Radosław author
Title: Enhancing modern living with virtual and augmented reality : practical applications / Kinga Stecuła and Radosław Wolniak.
Description: First edition. | Boca Raton, FL : CRC Press, 2026. | Series: CRC Press reference books in computer science | Includes bibliographical references and index.
Identifiers: LCCN 2026001792 (print) | LCCN 2026001793 (ebook) | ISBN 9781041080282 hbk | ISBN 9781041080305 pbk | ISBN 9781003643395 ebk
Subjects: LCSH: Augmented reality | Virtual reality
Classification: LCC QA76.9.A94 S74 2026 (print) | LCC QA76.9.A94 (ebook)
LC record available at https://lccn.loc.gov/2026001792
LC ebook record available at https://lccn.loc.gov/2026001793

ISBN: 978-1-041-08028-2 (hbk)
ISBN: 978-1-041-08030-5 (pbk)
ISBN: 978-1-003-64339-5 (ebk)

DOI: 10.1201/9781003643395

Typeset in Minion
by KnowledgeWorks Global Ltd.

Contents

Introduction, ix

CHAPTER 1 ▪ Virtual and Augmented Reality Concept 1

1.1 DEFINITION AND HISTORY OF VIRTUAL REALITY 1

1.2 DEFINITION AND HISTORY OF AUGMENTED REALITY 5

1.3 AREAS OF PRACTICAL APPLICATION OF VIRTUAL AND AUGMENTED REALITY 10

CHAPTER 2 ▪ Entertainment and Culture 18

2.1 GAMES, MOVIES, AND CULTURAL EVENTS IN VIRTUAL REALITY 18

2.2 AUGMENTED REALITY IN INTERACTIVE ENTERTAINMENT 23

2.3 IMMERSION EXPERIENCES IN VIRTUAL AND AUGMENTED REALITY 28

CHAPTER 3 ▪ Shopping and Retail 33

3.1 APPLICATIONS FOR SHOPPING IN VIRTUAL REALITY 33

3.2 AUGMENTED REALITY SOLUTIONS IN RETAIL 38

3.3 MARKETING IN THE CONTEXT OF VIRTUAL AND AUGMENTED REALITY 45

CHAPTER 4 ▪ Sport 51

4.1 VIRTUAL TRAININGS, EXERCISES, AND ACTIVITIES 51

4.2 FROM WATCHING TO PARTICIPATING IN SPORT THROUGH AUGMENTED REALITY 55

4.3 THE POTENTIAL OF VIRTUAL AND AUGMENTED REALITY TO ACTIVATE SOCIETY 59

CHAPTER 5 ■ Tourism 63

5.1 VIRTUAL TOURISM: EXPLORING THE WORLD WITHOUT LEAVING HOME 63

5.2 AUGMENTED REALITY ENHANCING TRAVEL EXPERIENCES 70

5.3 OPPORTUNITIES AND POSSIBILITIES OF VIRTUAL AND AUGMENTED REALITY IN TOURISM 74

CHAPTER 6 ■ Relaxation and Well-Being 80

6.1 VIRTUAL REALITY FOR RELAXATION AND STRESS RELIEF 80

6.2 AUGMENTED REALITY FOR CREATING CALMING AND STRESS-FREE SPACES 85

6.3 POTENTIAL OF VIRTUAL AND AUGMENTED REALITY TO ENHANCE WELL-BEING 90

CHAPTER 7 ■ Virtual and Augmented Reality Enhancing Modern Living 97

7.1 ENHANCING LIVING IN THE ANALYZED AREAS 97

7.2 GENERAL BENEFITS OF USING VIRTUAL AND AUGMENTED REALITY 101

7.3 VIRTUAL AND AUGMENTED REALITY IN THE CONTEXT OF SMART CITIES 107

CHAPTER 8 ■ Limitations and Threats of Virtual and Augmented Reality 118

8.1 IDENTIFYING LIMITATIONS 118

8.2 IDENTIFYING THREATS 122

8.3 ADDRESSING CHALLENGES AND PROPOSING SOLUTIONS 126

Chapter 9 ■ Future of Virtual and Augmented Reality 137

9.1 EMERGING TRENDS AND TECHNOLOGIES IN VIRTUAL AND AUGMENTED REALITY 137

9.2 INTEGRATION OF VIRTUAL AND AUGMENTED REALITY WITH OTHER EMERGING TECHNOLOGIES 142

9.3 ETHICAL AND SOCIETAL IMPLICATIONS OF FUTURE VIRTUAL AND AUGMENTED REALITY ADOPTION 145

INDEX, 155

Introduction

The modern world is developing rapidly, and one of the main visible changes is the digital transformation. Technological advances, especially in the area of digital technologies, have an impact on a more and more number of areas of life – from industry and education to everyday life. Many countries around the world actively implement digitalization solutions, striving to create modern, intelligent societies living in smart cities. In this context, immersive technologies such as virtual reality (VR) and augmented reality (AR) are gaining increasing importance. They enable partial or complete immersion in a virtual world or its elements. VR transports the user to a three-dimensional, fully virtual world, while AR puts virtual elements on the surrounding real world. VR and AR are finding a growing number of applications, from entertainment and education to professional training. These technologies can also be used in various ways in everyday life. The market offers more and more VR and AR applications and solutions that support people in their daily lives and improve the quality and comfort of their lives. Not only entertainment and culture should be mentioned, but also shopping, retail, sports, tourism, relaxation, and well-being.

The aim of this book is to demonstrate the role of virtual and augmented reality in the contemporary world and present their practical applications in various spheres of life. The authors analyse examples of the use of VR and AR in entertainment, culture, commerce, sports, tourism, as well as in the areas of relaxation and well-being. They show how these technologies can facilitate everyday life, support social and economic development, and contribute to improving the quality of human functioning in the digital age.

The basic research questions in this book include the following:

- What areas of application of virtual and augmented reality can help facilitate the everyday life of the user through the use of these technologies?

- What are the key applications of virtual and augmented reality to enhance modern living in the digital age?
- What are the solutions and applications related to virtual and augmented reality in entertainment and culture?
- In what ways can virtual and augmented reality support and enhance the shopping experience?
- How can society practice sports and increase physical activity using virtual and augmented reality?
- How can the tourism industry be digitally developed using virtual and augmented reality?
- How can virtual and augmented reality affect the relaxation and well-being of users?
- What are the upcoming trends and technologies related to virtual and augmented reality?
- How do immersive technologies integrate with emerging technologies such as artificial intelligence and the Internet of Things?
- What are the major limitations, threats, and challenges associated with VR and AR?
- How can VR and AR contribute to the development of smart cities and the enhancement of modern living?
- What are the ethical implications of the widespread adoption of virtual and augmented reality technologies?

The following chapters present both practical solutions and case studies that demonstrate the real possibilities of using immersive technologies in the future. The potential of VR and AR in the context of smart cities, sustainable development, and digital accessibility is also discussed. The book refers also to limitations, risks, and challenges of VR and AR. It addresses the impact of these technologies on the psyche of users, the ethics of their use, as well as issues related to data privacy, addiction, and digital exclusion. It also identifies directions of development that can ensure the safe and responsible use of these technologies' potential.

Kinga Stecuła
Radosław Wolniak

CHAPTER 1

Virtual and Augmented Reality Concept

1.1 DEFINITION AND HISTORY OF VIRTUAL REALITY

The modern world is developing at a rapid pace, and the boundaries between physical and digital reality are becoming increasingly blurred. Dynamic technological advances, particularly in virtual reality (VR) and augmented reality (AR), are opening up new ways for modern people to experience, learn, and create. Immersive technologies, which until recently seemed the domain of science fiction, are now becoming part of everyday life – from education and medicine to art, entertainment, and communication. Virtual reality, one of the pillars of this new digital world, allows users to become completely immersed in a computer-generated world, offering experiences that engage not only the senses but also the mind, emotions, and imagination.

Virtual reality technology uses computer technology to create the effect of a three-dimensional interactive world in which objects give the impression of being physically present (Biocca & Lanier, 1992; Lanier, 1992). This definition was developed by American computer scientist and futurologist Jaron Lanier – he is considered the creator of the term. He was the founder of VPL Research, which produced a prototype of VR equipment. VR simulates the real or presents fictional world, including the whole space, individual objects, or events (Sherman & Craig, 2003). A user can observe such a world, feel as if they are there, and actively interact with virtual elements. Using controllers or even users' own hands (when it comes to

DOI: 10.1201/9781003643395-1

some headsets), VR user can touch, drag and drop objects, and more. In other words, virtual reality is an alternative world composed of computer-generated images that respond to human movements (Steuer et al., 1995). A user can move into the virtual world using special VR goggles, which make them not see the real world around them, but have before their eyes an unreal space and images displayed using the goggles.

The origins of virtual reality technology date back to the mid-20th century, when the first attempts to simulate multiple senses and immerse the user in another world emerged (Norman, 2021b). One of the first such devices was the Sensorama, designed by Morton Heilig around 1957 and patented in 1962. Heilig, sometimes referred to as the 'father of virtual reality' strove to create what he described as a multi-sensory theatre – a system that would stimulate viewers not only with images and sounds, but also with wind, smells, and vibrations (Islami, 2020). The Sensorama displayed a video depicting, for example, a motorcycle ride through the streets of Brooklyn, and the device accompanied the experience with additional stimuli (Lebiedź, 2015):

- A stereoscopic 3D image on a screen in front of the user's eyes.
- Surround sound.
- Wind effects (fans simulating gusts).
- Seat movements/vibrations to reflect the sensation of a dynamic journey.
- An aroma diffuser – emitting scents consistent with the film's content, e.g., the smell of city exhaust or ambient odours.

As a next step, Morton Heilig developed a smaller version of the device – the Telesphere Mask – glasses with screens that displayed stereoscopic images and stereo sound (Gupta et al., 2020). However, this version did not yet feature head tracking.

In 1968, a pioneering device known as the Sword of Damocles was created by Ivan Sutherland (with the help of Bob Sproull, among others) (Boas, 2013). This device was the first in history to use a computer to generate images, rather than a camera. However, the device was very heavy and had to be suspended from the ceiling – hence the name (because it looked like a hanging structure above the user's head). The system tracked the user's head movements and, based on this, changed the perspective in the generated virtual world. Sutherland humorously nicknamed the device the 'Sword of Damocles' because of its appearance. Additionally, because the device used partially transparent elements (reflectors) to overlay computer-generated images onto the real world, it is also considered a forerunner of AR technology.

In the following decades, the development of VR technology was fuelled by various companies and commercial projects. One of the first companies to begin selling VR hardware was VPL Research (Gupta et al., 2020), although it struggled financially in the 1990s and was eventually acquired by Sun Microsystems.

In 1995, Forte Technologies unveiled the VFX1 Headgear – a headset with built-in headphones, dual 263x230-pixel displays, and capable of displaying up to 256 colors. Head movements were tracked by external sensors, and interaction with the VR experience was enabled by a mouse-like controller connected to a PC.

Virtual reality gained renewed attention thanks to the Oculus Rift project, which ushered in a new era of accessible and effective immersive technologies. In 2012, Palmer Luckey, a young enthusiast and computer hardware designer, raised funds to develop the first modern VR headset – the Oculus Rift Development Kit 1 – through the crowdfunding platform Kickstarter (Clark, 2014). The campaign was a resounding success and sparked a resurgence of the VR market. The device boasted parameters that represented significant technological advances at the time:

- It featured an LCD display with a resolution of 640 × 800 pixels per eye.
- An image refresh rate of 60 Hz, significantly reducing lag and improving user comfort.
- The field of view spanned approximately 90° vertically and 90° horizontally, creating a much more natural feeling of immersion in the virtual environment.

By carefully tuning optical parameters and minimizing lag (motion sickness), Oculus Rift became the first next-generation VR device to effectively reduce symptoms such as nausea and headaches. In 2014, Oculus VR was acquired by Meta (formerly Facebook), sparking the rapid growth of the consumer VR market and the creation of numerous subsequent headset models (including the Oculus Rift S from 2019, Oculus Quest 2 from 2020, Meta Quest 3S from 2024, and more).

In 2016, HTC successfully launched its VR headset, the HTC Vive. One of the key innovations was support for six-degree-of-freedom (6DOF) motion tracking, meaning the user can move in three dimensions (forward-backward, up-down, side-to-side) and rotate their head and body in all directions (HTC, 2023). The Vive also supports full room-scale, meaning the user's movement in real space is tracked and replicated in the virtual environment – the user can move around a designated room, and the system responds to those movements. This company has also released more sets of goggles, among others, HTC Vive Pro in 2018, HTC Vive Cosmos Eliite in 2020, HTC Vive XR Elite in 2023, HTC Vive Focus Vision in 2024.

Currently, there are many headsets available for users all over the world. Despite the mentioned, the most popular headsets include the following: Valve Index HDM, Windows Mixed Reality, Sony PlayStation VR, Pico Neo, and many more (Valve Corporation, 2022). Figure 1.1 shows an illustrative drawing of a person with virtual reality goggles.

FIGURE 1.1 An illustrative drawing of a person with virtual reality goggles.

A virtual environment allows for so-called immersion within it. The concept of immersion is crucial in the context of virtual reality and is inextricably linked to it. Immersion is a metaphorical term derived from the physical experience of being immersed in water (Murray, 2017). The term is also used to refer to experiences that engage a person psychologically. It is a phenomenon experienced by an individual when they are in a state of deep mental engagement, in which their cognitive processes (with or without sensory stimulation) cause a shift in their state of attention such that they experience a disconnection from awareness of the physical world (Agrawal et al., 2019). A video game can be immersive. Playing an immersive game allows a person to forget about the real world. Music can also be immersive. Listening to touching music, a person can lose themselves, forgetting the world around them. In other words, it is the feeling a person experiences when immersed in the ocean or a swimming pool: the sensation of being surrounded by a completely different reality (different as much as water differs from air) that takes over his entire attention, his entire perceptual apparatus.

1.2 DEFINITION AND HISTORY OF AUGMENTED REALITY

AR involves superimposing digital information onto the physical world, coinciding with it both spatially and temporally, and with which the user can interact in real time (Craig, 2013). Thus, AR bridges the real and virtual worlds. Visualizations, most often 3D, are superimposed on the real world. AR can therefore be said to be an intermediate form between the virtual and real worlds. Unlike VR, AR does not create a completely new virtual world, but adds content. It overlays images or information onto the real world, which remains unchanged (Dejnaka, 2012). This technology was introduced chronologically later than virtual reality (Berbeka, 2016). Virtual elements can take many forms. They can be simple display information, such as the name of an item, which appears when the user looks at it. The second type are visual objects. They integrate with the real world and create a coherent whole. An example is a virtual piece of furniture that we see when looking at a real room in which we are located. In conjunction with the displayed information, using AR in a city, it can include street names, monuments, and stops (Wieczorek & Stecuła, 2025). In a museum, a description of a given painting (information about the creator, title, painting style) can be displayed. In a body shop, a mechanic using AR can see the name of a given part.

AR can take many forms, depending on how the user perceives the connection between the real world and virtual elements. Currently, there are many techniques for presenting AR content. The most important forms of AR are listed below:

- Wearable AR: Uses special goggles with transparent displays that allow the user to see the real world enhanced with digital objects. This type allows for complete freedom of movement (for example, Microsoft HoloLens, Magic Leap, Google Glass).
- Mobile AR: Relies on smartphones or tablets that use a built-in camera to capture the surroundings and overlay virtual elements on them (for example, playing Pokémon GO application, or using shopping application like IKEA Place).
- Head-Up Displays (HUDs): Used primarily in aviation, automotive, and industrial applications. Information is displayed directly on a transparent surface, such as a car windshield or the pilot's visor (for example, head-up display systems in cars, flight helmets with information visors).
- Desktop AR: A less common form of AR in which a computer camera captures a real image and superimposed virtual objects are displayed on the screen. Used primarily in education, design, and scientific research (for example, marker systems for learning anatomy, design programs with 3D previews).
- Projection-Based AR: In this form, virtual images are projected onto real surfaces using projectors. The user does not need any goggles, and the image is visible to the naked eye. Such solutions are used in museums, interactive exhibitions, and in industry (for example, projectors presenting assembly instructions).
- Spatial AR combines various technologies (depth sensors, cameras, projectors), allowing the system to 'understand' the user's space and embed virtual objects in the real environment. This allows for the illusion of the physical presence of digital objects in the real world (for example, interactive exhibitions, architectural presentations, interior design systems).

Figure 1.2 presents an illustrative drawing of a person with AR goggles who sees a virtual chair in a room.

FIGURE 1.2 A person with AR goggles looking at a virtual chair with its description.

AR technology functions in such a way that it recognizes objects in the real world and superimposes virtual elements on them (Lee et al., 2022). Using AR involves putting on special goggles on the head that do not cover the field of view and through which information or virtual images are displayed. In addition, the user uses controllers that allow the user to manipulate virtual objects (Huo et al., 2021). Although in practice, in the case of AR, the technology is so advanced that the user uses his own hands (Grodotzki et al., 2023; He & Yang, 2014) without any additional sensors. Another solution is a special glove (Minh et al., 2019), thanks to which the AR system recognizes gestures, as well as other wearable haptics solutions (Maisto et al., 2017). Therefore, AR technology not only tracks and detects the movement of the head (goggles) as well as the movement of the eyeballs (Kapp et al., 2021), but also recognizes direct hand gestures and even individual fingers (Liebers et al., 2024). Hence, it is about immersion (Dudley et al., 2023), that is, immersion in a new world. Figure 1.3 shows the view from the AR goggles: the user sees a real room and a widget suspended in the air.

Currently, AR is intended for a single user, although it may develop towards the so-called 'multiplayer'. This is the case with VR, for which solutions allow several users to use the same application (Karaosmanoglu et al., 2021;

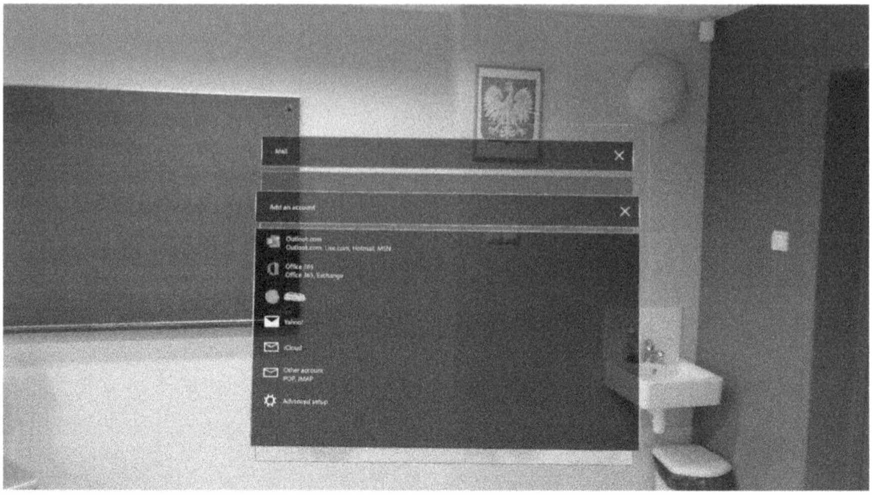

FIGURE 1.3 A view from AR goggles: a real room and a virtual widget.

Rogers et al., 2021). What is more, there is also asymmetric VR (Stecuła, 2024), which allows for the joint use of a given application (or game) by different users who have different devices (for example, one has VR goggles and the other plays in front of a computer – and they play together).

It is worth adding that AR is one of the pillars of Industry 4.0 which is increasingly present in the functioning of organizations such as enterprises (De Pace et al., 2018). It is assumed that AR will revolutionize production and operational processes by integrating digital information with the physical world. AR allows employees to access assembly instructions and diagnostics in real time. It also supports training, allowing for simulations and visualizations. Thanks to AR, it is possible to monitor and optimize complex production processes. However, AR offers great opportunities for use in various industries and areas.

VR and AR technologies therefore differ significantly. The goal of VR is to create the illusion of the real world or a fragment of it, where the user will feel as if they have been transported to another reality, separated from the physical world around them. VR goggles obscure the user's field of vision and present three-dimensional images and emit sounds. AR, on the other hand, does not create a virtual world; rather, it recognizes real-world objects and superimposes virtual elements on them.

Examples of VR applications include Half-Life: Alyx, Walkabout Mini Golf VR, The Room VR: A Dark Matter, and Beat Saber – games that were highly evaluated on the Steam platform by players worldwide (Stecuła, 2022).

An example of AR is applying Snapchat or Instagram filters applied to a user's face. Using a camera and looking at objects, users can add, for example, virtual cat ears to their own face. Another example is Google Maps Live View (Barbu, 2019). The app can be downloaded to a phone, and when looking through it at the real world, users see, for example, superimposed virtual arrows. These are intended to help a user reach their destination.

The origins of AR technology date back to the 1960s. The Sword of Damocles, created in 1968 by Ivan Sutherland at Harvard University, is considered one of the first devices to combine the real world with computer-generated elements (Boas, 2013). This system was a forerunner of both virtual and augmented reality. Using partially transparent screens, the user could simultaneously see the real world and computer-generated images superimposed on it. For this reason, Sutherland's project is considered a precursor to modern AR technology.

The next important step in the development of AR occurred in 1992, when researchers Tom Caudell and David Mizell of Boeing Aerospace first coined the term 'augmented reality' (Shubham, 2023). While working to improve aircraft assembly, they developed a system using a head-mounted display that presented computer-generated diagrams and assembly instructions directly to technicians in their field of view. The new solution reduced production costs and increased work efficiency by eliminating the need for paper documents, plans, and diagrams. AR thus found practical application in industry for the first time, becoming a tool that supports humans in performing complex technical tasks.

Another breakthrough in the history of AR came in 1992, when Louis Rosenberg of the US Armstrong Research Lab developed the Virtual Fixtures system (Norman, 2021a). It was one of the first fully functional AR systems developed for military use. The device assisted pilots during training and simulations by superimposing virtual elements onto the real field of view, acting as 'assistive overlays' – facilitating spatial orientation, maneuvering, and interaction with the simulator. Rosenberg's project was a pioneering step towards the practical use of AR as a tool for increasing the efficiency and safety of operations in military environments.

In 1996, Japanese researcher Jun Rekimoto presented a prototype device called the NaviCam, which combined video camera images with real-time computer information (Rekimoto, 1997). The NaviCam allowed the user to point to objects in the environment and obtain additional information about them. This was one of the first examples of a mobile AR system to utilize the idea of augmenting the real world with contextually accessible digital data.

In 1997, scientist Ronald Azuma published a groundbreaking article titled 'A Survey of Augmented Reality', in which he introduced and systematized the concept of AR. Many researchers consider Azuma to have created the first definition of AR, still in use today. According to his concept, AR systems are characterized by three fundamental characteristics (Azuma, 1997):

- Combining the Real and Virtual Worlds: The starting point is the user's actual environment, overlaid with digital elements.
- Real-Time Interactivity: The system responds immediately to the user's actions, creating the impression of both worlds coexisting.
- Three-Dimensional Projection: The user can observe virtual elements from different perspectives and in motion, lending them dynamics and realism.

The definition proposed by Azuma remains a reference point in the scientific and engineering literature on AR technology to this day.

1.3 AREAS OF PRACTICAL APPLICATION OF VIRTUAL AND AUGMENTED REALITY

Both virtual and augmented reality are widely used in various areas of human activity. In literature and practice, there are many examples of applications used for various purposes, for example, training in various industries.

As it can be read in different scientific papers, VR can be applied in the following areas: medicine (Li et al., 2017), aviation (Fussell, 2020), military (Pallavicini et al., 2016), architecture (Bashabsheh et al., 2019), chemistry (Fracaro et al., 2021), and automotive industry (Lawson et al., 2016). It finds also a practical usage in psychology (Parsons et al., 2017), rehabilitation (Smits et al., 2020), healthcare (Izard et al., 2017) but also it can be used in education (Soliman et al., 2021), tourism (Beck et al., 2019; Stecuła & Naramski, 2025), marketing (Ozdemir, 2021), and many others. There are different purposes of using VR, for example: to be useful in different therapies (Caponnetto et al., 2021), to overcome depression (Baghaei et al., 2021),

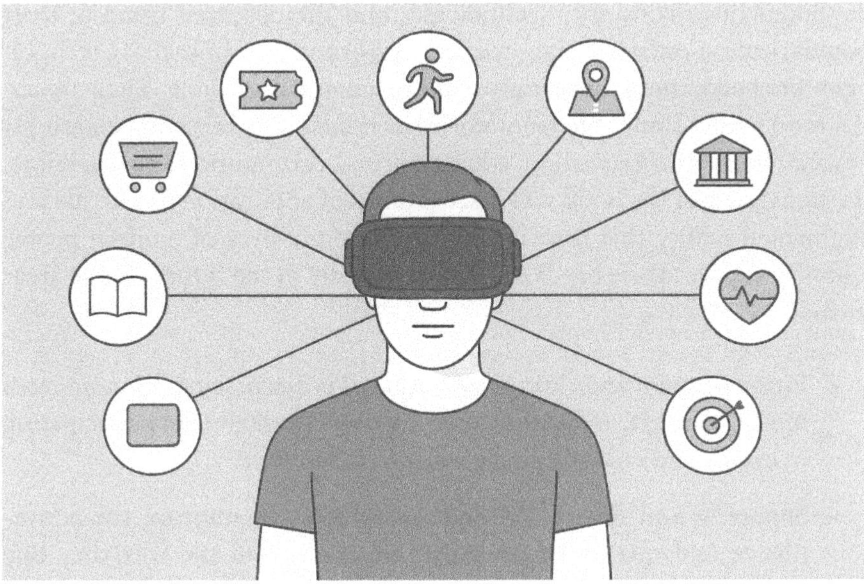

FIGURE 1.4 An illustrative drawing of a different usage of VR.

to reduce stress (Yang et al., 2021), in surgeries (Hsiao et al., 2021), driving simulations (Saito et al., 2020), autonomous vehicles (Dou et al., 2021), assemblies (Roldán et al., 2019), monitoring and safety systems (Alam et al., 2017), marine conservation (Yuen et al., 2023), and more.

Figure 1.4 is an illustrative drawing which presents symbols of a different usage of VR.

In turn, AR is not currently used on a global scale, but its potential is huge. AR technology can be used in entertainment (Bueno et al., 2020; Hung et al., 2021), sport (Sawan et al., 2020), tourism and culture (concerts, museums, sightseeing) (Ozdemir, 2021; Wang & Zhu, 2022; Zhu & Wang, 2022). There are also examples of its usage in medicine (Yeung et al., 2021), education (Lu et al., 2021), architecture (Hajirasouli & Banihashemi, 2022), production (Hořejší et al., 2020), shopping (Cruz et al., 2019; Wolniak et al., 2024), trade and advertising (Bonetti et al., 2018; Zhu & Wang, 2022), treatment and therapies (Doulou & Drigas, 2022), and many others.

There are currently many areas of application for VR and AR. This book, however, focuses on everyday applications – those that facilitate human functioning, impact quality of life, and may in the future become an integral part of everyday reality. This choice is dictated by the dynamic development of consumer technologies and the observed trend of their popularization outside the professional environment. Immersive technologies are becoming increasingly widespread and accessible. Their potential is no longer limited to laboratories or industry – they are increasingly becoming tools for recreation, education, and communication. Therefore, the remainder of the book will discuss areas of application for virtual and augmented reality that directly impact the daily lives of modern people and may support their needs even more broadly in the future. These areas include:

- Entertainment and Culture: VR and AR is discussed in the context of music, film, art, and games, offering users the feeling of participating in events and creating a new way of interactivity.
- Shopping and Retail: VR and AR which can improve the convenience and quality of consumer decisions and the way they buy things.
- Sports: VR and AR are used for training, movement analysis, physical simulation, and rehabilitation support.
- Tourism: VR and AR enable virtual visits, travel planning, and enriching tourist experiences through interactive content and historical visualizations.
- Relaxation and Well-Being: VR and AR can support relaxation, meditation, and stress management, allowing for transport to peaceful, digital spaces conducive to mental regeneration.

The next five chapters describe applications and solutions that are used and have great potential for future use in the areas mentioned above.

BIBLIOGRAPHY

Agrawal, S., Simon, A., Bech, S., Bærentsen, K., & Forchhammer, S. (2019). Defining immersion: Literature review and implications for research on immersive audiovisual experiences. *Journal of Audio Engineering Society, 68*(6), 404–417.

Alam, M. F., Katsikas, S., Beltramello, O., & Hadjiefthymiades, S. (2017). Augmented and virtual reality based monitoring and safety system: A prototype IoT platform. *Journal of Network and Computer Applications, 89*, 109–119. https://doi.org/10.1016/j.jnca.2017.03.022

Azuma, R. T. (1997). A survey of augmented reality. *Presence: Teleoperators & Virtual Environments*, *6*(4), 355–385.

Baghaei, N., Chitale, V., Hlasnik, A., Stemmet, L., Liang, H.-N., & Porter, R. (2021). Virtual reality for supporting the treatment of depression and anxiety: Scoping review. *JMIR Mental Health*, *8*(9), e29681.

Barbu, P. (2019). *Google Maps introduces AR walking directions in new Live View feature*. https://business-review.eu/tech/live-view-google-maps-203857

Bashabsheh, A. K., Alzoubi, H. H., & Ali, M. Z. (2019). The application of virtual reality technology in architectural pedagogy for building constructions. *Alexandria Engineering Journal*, *58*(2), 713–723.

Beck, J., Rainoldi, M., & Egger, R. (2019). Virtual reality in tourism: A state-of-the-art review. *Tourism Review*, *74*(3), 586–612.

Berbeka, J. (2016). Wirtualna i rozszerzona rzeczywistość a zachowania konsumentów. *Studia Ekonomiczne*, *303*, 84–101.

Biocca, F., & Lanier, J. (1992). An insider's view of the future of virtual reality. *Journal of Communication*, *42*(2), 156–157.

Boas, Y. (2013). Overview of virtual reality technologies. *Interactive Multimedia Conference*, *2013*, 4. https://www.semanticscholar.org/paper/Overview-of-Virtual-Reality-Technologies-Boas/4214cb09e29795f5363e5e3b545750dce027b668

Bonetti, F., Warnaby, G., & Quinn, L. (2018). Augmented reality and virtual reality in physical and online retailing: A review, synthesis and research agenda. *Augmented Reality and Virtual Reality*, 119–132. https://link.springer.com/chapter/10.1007/978-3-319-64027-3_9

Bueno, S., Gallego, M. D., & Noyes, J. (2020). Uses and gratifications on augmented reality games: An examination of pokémon go. *Applied Sciences*, *10*(5), 1644.

Caponnetto, P., Triscari, S., Maglia, M., & Quattropani, M. C. (2021). The simulation game—Virtual reality therapy for the treatment of social anxiety disorder: A systematic review. *International Journal of Environmental Research and Public Health*, *18*(24), 13209.

Clark, T. (2014). *How Palmer Luckey Created Oculus Rift*. https://www.smithsonianmag.com/innovation/how-palmer-luckey-created-oculus-rift-180953049/

Craig, A. B. (2013). *Understanding augmented reality: Concepts and applications*. Morgan Kaufmann.

Cruz, E., Orts-Escolano, S., Gomez-Donoso, F., Rizo, C., Rangel, J. C., Mora, H., & Cazorla, M. (2019). An augmented reality application for improving shopping experience in large retail stores. *Virtual Reality*, *23*, 281–291.

Dejnaka, A. (2012). Rzeczywistość rozszerzona i jej zastosowanie w edukacji. *E-Mentor*, *44*(2), 30–36.

De Pace, F., Manuri, F., & Sanna, A. (2018). Augmented reality in industry 4.0. *American Journal of Computer Science and Information Technology*, *6*(1), 17.

Dou, J., Chen, S., Tang, Z., Xu, C., & Xue, C. (2021). Evaluation of multimodal external human–machine interface for driverless vehicles in virtual reality. *Symmetry*, *13*(4), 687.

Doulou, A., & Drigas, A. (2022). Electronic, VR & augmented reality games for intervention in ADHD. *Technium Social Sciences Journal, 28*, 159.

Dudley, J., Yin, L., Garaj, V., & Kristensson, P. O. (2023). Inclusive immersion: A review of efforts to improve accessibility in virtual reality, augmented reality and the metaverse. *Virtual Reality, 27*(4), 2989–3020.

Fracaro, S. G., Chan, P., Gallagher, T., Tehreem, Y., Toyoda, R., Bernaerts, K., Glassey, J., Pfeiffer, T., Slof, B., & Wachsmuth, S. (2021). Towards design guidelines for virtual reality training for the chemical industry. *Education for Chemical Engineers, 36*, 12–23.

Fussell, S. G. (2020). *Determinants of aviation students' intentions to use virtual reality for flight training.* Embry-Riddle Aeronautical University.

Grodotzki, J., Müller, B. T., & Tekkaya, A. E. (2023). Enhancing manufacturing education based on controller-free augmented reality learning. *Manufacturing Letters, 35*, 1246–1254.

Gupta, R., Kumar, K., & Shah, S. D. (2020). *A Brief History of Virtual Reality, in Patents.* https://www.finnegan.com/en/insights/articles/a-brief-history-of-virtual-reality-in-patents.html

Hajirasouli, A., & Banihashemi, S. (2022). Augmented reality in architecture and construction education: State of the field and opportunities. *International Journal of Educational Technology in Higher Education, 19*(1), 39.

He, Z., & Yang, X. (2014). Hand-based interaction for object manipulation with augmented reality glasses. *Proceedings of the 13th ACM SIGGRAPH International Conference on Virtual-Reality Continuum and Its Applications in Industry*, 227–230.

Hořejší, P., Novikov, K., & Šimon, M. (2020). A smart factory in a smart city: Virtual and augmented reality in a smart assembly line. *IEEE Access, 8*, 94330–94340.

Hsiao, Y.-C., Chang, C.-J., & Fang, J.-J. (2021). Quantitative asymmetry assessment between virtual and mixed reality planning for orthognathic surgery—A retrospective study. *Symmetry, 13*(9), 1614.

HTC. (2023). *HTC.* https://www.vive.com

Hung, S.-W., Chang, C.-W., & Ma, Y.-C. (2021). A new reality: Exploring continuance intention to use mobile augmented reality for entertainment purposes. *Technology in Society, 67*, 101757.

Huo, J., Keung, K. L., Lee, C. K. M., & Ng, H. Y. (2021). Hand gesture recognition with augmented reality and leap motion controller. *2021 IEEE International Conference on Industrial Engineering and Engineering Management (IEEM)*, 1015–1019.

Islami, D. (2020). *VIRTUAL REALITY - VR* (Master's thesis). University for Business and Technology. https://knowledgecenter.ubt-uni.net/etd/1744

Izard, S. G., Juanes Méndez, J. A., & Palomera, P. R. (2017). Virtual reality educational tool for human anatomy. *Journal of Medical Systems, 41*, 1–6.

Kapp, S., Barz, M., Mukhametov, S., Sonntag, D., & Kuhn, J. (2021). ARETT: Augmented reality eye tracking toolkit for head mounted displays. *Sensors, 21*(6), 2234.

Karaosmanoglu, S., Rogers, K., Wolf, D., Rukzio, E., Steinicke, F., & Nacke, L. E. (2021). Feels like team spirit: Biometric and strategic interdependence in asymmetric multiplayer VR games. *Proceedings of the 2021 CHI Conference on Human Factors in Computing Systems*, 1–15.

Lanier, J. (1992). Virtual reality: The promise of the future. *Interactive Learning International, 8*(4), 275–279.

Lawson, G., Salanitri, D., & Waterfield, B. (2016). Future directions for the development of virtual reality within an automotive manufacturer. *Applied Ergonomics, 53*, 323–330.

Lebiedź, J. (2015). *Człowiek zanurzony w rzeczywistości wirtualnej na przykładzie Laboratorium Zanurzonej Wizualizacji Przestrzennej.* https://mostwiedzy.pl/pl/publication/czlowiek-zanurzony-w-rzeczywistosci-wirtualnej-na-przykladzie-laboratorium-zanurzonej-wizualizacji-p%2C135003-1?utm

Lee, T., Jung, C., Lee, K., & Seo, S. (2022). A study on recognizing multi-real world object and estimating 3D position in augmented reality. *The Journal of Supercomputing, 78*(5), 7509–7528.

Li, L., Yu, F., Shi, D., Shi, J., Tian, Z., Yang, J., Wang, X., & Jiang, Q. (2017). Application of virtual reality technology in clinical medicine. *American Journal of Translational Research, 9*(9), 3867.

Liebers, J., Brockel, S., Gruenefeld, U., & Schneegass, S. (2024). Identifying users by their hand tracking data in augmented and virtual reality. *International Journal of Human–Computer Interaction, 40*(2), 409–424.

Lu, A., Wong, C. S. K., Cheung, R. Y. H., & Im, T. S. W. (2021). Supporting flipped and gamified learning with augmented reality in higher education. *Frontiers in Education, 6*, 110.

Maisto, M., Pacchierotti, C., Chinello, F., Salvietti, G., De Luca, A., & Prattichizzo, D. (2017). Evaluation of wearable haptic systems for the fingers in augmented reality applications. *IEEE Transactions on Haptics, 10*(4), 511–522.

Minh, V. T., Katushin, N., & Pumwa, J. (2019). Motion tracking glove for augmented reality and virtual reality. *Paladyn, Journal of Behavioral Robotics, 10*(1), 160–166.

Murray, J. H. (2017). *Hamlet on the holodeck, updated edition: The future of narrative in cyberspace*. MIT press.

Norman, J. (2021a). *Louis Rosenberg Develops Virtual Fixtures, the First Fully Immersive Augmented Reality System.* https://www.historyofinformation.com/detail.php?id=4231

Norman, J. (2021b). *The Sensorama: One of the First Functioning Efforts in Virtual Reality.* https://www.historyofinformation.com/detail.php?id=2785&utm

Ozdemir, M. A. (2021). Virtual reality (VR) and augmented reality (AR) technologies for accessibility and marketing in the tourism industry. In: *ICT tools and applications for accessible tourism* (pp. 277–301). IGI Global.

Pallavicini, F., Argenton, L., Toniazzi, N., Aceti, L., & Mantovani, F. (2016). Virtual reality applications for stress management training in the military. *Aerospace Medicine and Human Performance, 87*(12), 1021–1030.

Parsons, T. D., Riva, G., Parsons, S., Mantovani, F., Newbutt, N., Lin, L., Venturini, E., & Hall, T. (2017). Virtual reality in pediatric psychology. *Pediatrics, 140*(Supplement_2), S86–S91.

Rekimoto, J. (1997). NaviCam: A magnifying glass approach to augmented reality. *Presence: Teleoperators and Virtual Environments*, 6(4), 399–412. https://doi.org/10.1162/pres.1997.6.4.399

Rogers, K., Karaosmanoglu, S., Wolf, D., Steinicke, F., & Nacke, L. E. (2021). A best-fit framework and systematic review of asymmetric gameplay in multiplayer virtual reality games. *Frontiers in Virtual Reality, 2*, 694660.

Roldán, J. J., Crespo, E., Martín-Barrio, A., Peña-Tapia, E., & Barrientos, A. (2019). A training system for industry 4.0 operators in complex assemblies based on virtual reality and process mining. *Robotics and Computer-Integrated Manufacturing, 59*, 305–316.

Saito, Y., Kawashima, K., & Hirakawa, M. (2020). Effectiveness of a head movement interface for steering a vehicle in a virtual reality driving simulation. *Symmetry, 12*(10), 1645.

Sawan, N., Eltweri, A., De Lucia, C., Pio Leonardo Cavaliere, L., Faccia, A., & Roxana Moşteanu, N. (2020). Mixed and augmented reality applications in the sport industry. *Proceedings of the 2020 2nd International Conference on E-Business and E-Commerce Engineering*, 55–59.

Sherman, W. R., & Craig, A. B. (2003). *Understanding virtual reality. Morgan Kauffman.*

Shubham, P. (2023). *Augmented Reality Over The Years*. https://www.agilis.no/en/post/augmented-reality-over-the-years

Smits, M., Staal, J. B., & Van Goor, H. (2020). Could virtual reality play a role in the rehabilitation after COVID-19 infection? *BMJ Open Sport & Exercise Medicine*, 6(1), e000943.

Soliman, M., Pesyridis, A., Dalaymani-Zad, D., Gronfula, M., & Kourmpetis, M. (2021). The application of virtual reality in engineering education. *Applied Sciences, 11*(6), 2879.

Stecuła, K. (2022). Virtual reality applications market analysis—On the example of steam digital platform. *Informatics*, 9(4), 100.

Stecuła, K. (2024). Analysis of asymmetric VR games—Steam platform case study. *Technology in Society, 78*, 102673. https://doi.org/10.1016/j.techsoc.2024.102673

Stecuła, K., & Naramski, M. (2025). Virtual reality as a green tourism alternative: Social acceptance and perception. *Sustainability, 17*(17), 7722.

Steuer, J., Biocca, F., & Levy, M. R. (1995). Defining virtual reality: Dimensions determining telepresence. *Communication in the Age of Virtual Reality, 33*, 37–39.

Valve Corporation. (2022). *Steampowered*. https://steampowered.com

Wang, C., & Zhu, Y. (2022). A survey of museum applied research based on mobile augmented reality. *Computational Intelligence and Neuroscience, 2022*(1), 2926241.

Wieczorek, A., & Stecuła, K. (2025). Supporting sustainable demographic change: Augmented reality in Men's sheds as a concept for active aging. *Sustainability, 17*(18), 8345.

Wolniak, R., Stecuła, K., & Aydın, B. (2024). Digital transformation of grocery in-store shopping-scanners, artificial intelligence, augmented reality and beyond: A review. *Foods, 13*(18). https://doi.org/10.3390/foods13182948

Yang, T., Lai, I. K. W., Fan, Z. B., & Mo, Q. M. (2021). The impact of a 360 virtual tour on the reduction of psychological stress caused by COVID-19. *Technology in Society, 64*, 101514.

Yeung, A. W. K., Tosevska, A., Klager, E., Eibensteiner, F., Laxar, D., Stoyanov, J., Glisic, M., Zeiner, S., Kulnik, S. T., & Crutzen, R. (2021). Virtual and augmented reality applications in medicine: Analysis of the scientific literature. *Journal of Medical Internet Research, 23*(2), e25499.

Yuen, K. F., Koh, L. Y., Tan, L. Y. H., & Wang, X. (2023). The determinants of virtual reality adoption for marine conservation. *Technology in Society, 72*, 102170.

Zhu, Y., & Wang, C. (2022). Study on virtual experience marketing model based on augmented reality: Museum marketing (example). *Computational Intelligence and Neuroscience, 2022*(1), 2485460.

CHAPTER 2

Entertainment and Culture

2.1 GAMES, MOVIES, AND CULTURAL EVENTS IN VIRTUAL REALITY

Virtual reality (VR) is widely used in the entertainment and cultural industries. It is in this area that this technology is developing most dynamically, as it allows users to enter a completely new digital world. Using VR headsets, controllers, and appropriate applications, users can become completely immersed in a virtual environment – playing games, watching movies, attending concerts, or visiting museums without leaving home. VR is therefore becoming not only a form of entertainment but also an educational and cultural tool.

Games are one of the most important areas of VR development. Like traditional computer games, VR games are divided into many genres. These can include action, adventure, casual, simulation, strategy, puzzle, fantasy, exploration, music, sports, educational, horror, and more (Stecuła, 2022):

In the action genre, well-known titles include 'Superhot VR', 'Half-Life: Alyx', and 'The Walking Dead: Saints & Sinners'. Example of a popular (highly rated on Steam) music game is 'Beat Saber', in which players must rhythmically slice cubes with virtual lightsabers to the beat of the music. The game also provides good physical exercise, allowing players to tire themselves out and train muscle. Examples of VR game genres and their titles are presented in Table 2.1.

DOI: 10.1201/9781003643395-2

TABLE 2.1 Examples of VR Game Genres and Their Titles

VR Game Genre	Example of VR Game	Game Description
Action	Superhot VR	A game where time moves only when the player moves. It lets player feel like the hero of an action movie.
Adventure	Red Matter 2	The player takes on the role of an astronaut sent to an abandoned base. The game combines exploration, puzzle-solving, and elements of a science fiction thriller.
Casual	Job Simulator	A humorous game in which the player performs simple everyday tasks, such as cooking or working in the office.
Simulation	VTOL VR	A jet pilot simulator that lets the player control your plane with hand movements.
Strategy	Final Assault	A real-time strategy game (RTS) where the player command an army from a bird's-eye view in VR.
Puzzle	The Room VR: A Dark Matter	A puzzle game where the player solves complex puzzles in mysterious locations.
Fantasy	The Mage's Tale	The player takes on the role of a young mage who casts spells and battles monsters in a magical world.
Exploration	The Lab	A collection of mini-games and VR experiences, allowing the player to explore various virtual worlds.
Music	Beat Saber	A dynamic rhythm game where the player must rhythmically slice light cubes with virtual lightsabers to the beat of the music.
Sports	Eleven Table Tennis	A table tennis simulator with online play.
Educational	Titans of Space PLUS	An educational journey through the solar system, allowing the player to see the planets up close.
Horror	Phasmophobia VR	A cooperative game in which players explore haunted locations using ghost-detecting equipment.

Source: Author's own work.

VR games are available on numerous platforms, such as SteamVR, the Oculus Store (Meta Quest), Viveport, the PlayStation VR Store, and the Pico Store. There are also independent stores where developers publish their own titles. VR games are available in both single-player and multiplayer versions, meaning you can play alone or with other users from around the world. A growing number of companies are also creating dedicated, custom applications, such as educational VR games for schools, training simulators for athletes, and interactive marketing games for companies. Another interesting trend is asymmetric games, where one player uses VR goggles while the other plays, for example, on a computer or phone. This gameplay model blends the real and virtual worlds and allows for shared play. The example of such game is 'Keep Talking and Nobody Explodes'. Participants find themselves in an enclosed space with a virtual ticking time bomb, the basic instructions required to deactivate which are held by a second player. One player can see the bomb, while the other (the so-called expert) cannot see any visual representation of the explosive device (Steampowered, 2015). However, they do have instructions to defuse the bomb. The gameplay forces players to communicate quickly and effectively. Players next to the bomb can only see it and can only listen and talk to the expert. This game was not originally designed for VR, but a VR module was later added. In the literature, research on the asymmetric nature of some games can also be found. For example, Stecuła characterizes the asymmetric nature of asymmetric VR in the following way (Stecuła, 2024):

- Participants use different appliances – VR and non-VR devices (laptops, PCs, tablets, etc.).
- Participants experience gameplay from different viewpoints and interact with it in various ways – a person in VR experiences a 3D immersive environment and manipulates objects using controllers or hands; a person with a non-VR device is not in such immersive environment and uses different devices like a mouse or keyboard.
- Participants play different roles (characters), and have different responsibilities.
- Participants have different goals that contribute to the overall.
- Success of the gameplay.
- Participants perform different tasks and actions.
- Participants have access to different levels of information and resources.

Goggles alone are not enough to fully experience VR. On the modern market, there is a variety of devices that enhance the feeling of realism and immersion. One of the most interesting solutions is VR treadmills. They are special platforms that allow users to walk, run, or turn in any direction.

The platforms are physical but they let the user to move in the virtual world. Examples of such devices include the Virtuix Omni (Virtuix, 2025) and the Kat Walk C2 (KAT VR, 2025). This allows players to move in the game, instead of just stand still and move virtually by clicking the buttons on the controllers.

Another type of device is haptic gloves, for example, HaptX (Haptx, 2024) or SenseGlove (Senseglove, 2025), which provide the user with tactile stimuli. Thanks to these gloves, a user have a sense of holding objects, touch, or resistance. For example, when they grip a virtual sword or steer wheel, they can feel its structure. Haptic vests (e.g., bHaptics Tactsuit) are also becoming increasingly popular. They allow users to feel vibrations and movements in the game.

There are also cycling, kayaking, motorcycle, and climbing simulators. In these types of VR applications, the user uses a device that resembles real equipment – for example, pedaling on a stationary bike while viewing a mountain trail or city through the goggles. Examples include VirZOOM Bike (VIRZOOM, 2024) and the Holodia HOLOFIT kayak simulator (Holodia, 2025), which allows paddling on virtual rivers and lakes. Climbing games like The Climb 2, on the other hand, allow users to experience height and movement, thanks to sensors and vibrations. Such solutions combine sport, technology, and entertainment, while also allowing users to maintain their fitness in an engaging way.

VR is also increasingly being used in cinematography. Virtual reality films allow viewers to be 'inside' the action – they can look around, observe the world from the protagonist's perspective, and sometimes even influence the development of the plot. Many such films are presented as 360-degree videos, which can be viewed using VR headsets, for example, in apps like YouTube VR, Oculus TV, Within, or MetaQuest Video.

Interactive VR films allow users to make decisions – for example, choosing what the protagonist will do next. One pioneering example of this type of production is 'Henry' (produced by Oculus Story Studio), as well as 'Dear Angelica', which combines painterly aesthetics with emotional narrative. VR films are also being used in museums and galleries; for example, the Louvre is experimenting with 360-degree projections that transport viewers to the heart of works of art.

VR is also entering the world of culture and music. Users can experience live concerts without leaving home. An example is the WaveXR platform, which broadcasts virtual performances by artists in immersive 3D environments. As stated on WaveXR's website (Wave, 2025), its vision is to be the centre of live entertainment in the metaverse. It creates a place where anyone can participate, perform live, or be part of a 24/7 entertainment community. Users join on mobile devices to watch a 2D social video stream or immerse themselves in 3D VR. Virtual shows ('Waves') are at the most important part of the platform. Waves are live. They enable real-time (on-line) interaction and connection between performers and audience (Wave, 2025). Similar service is offered by application named Stageverse, which allow fans to meet in virtual clubs and music festivals.

In the context of cultural events, it's worth mentioning projects like Venues (Meta), where you can attend theatre performances, stand-up comedy shows, and conferences. Virtual events enable interaction with other participants and a sense of community, even though everyone is located in different parts of the world. Film festivals also try to use VR. The example can be Venice VR Expanded in Venice (La Biennale di Venezia, 2023) which offers audiences special sections for VR films and experiences.

It is worth mentioning applications on virtual tourism. They allow visitors to visit places that are difficult to reach in real life – such as mountain peaks, ocean floors, or monuments in distant countries (Stecuła & Naramski, 2025). Thanks to applications such as Google Earth VR (Steampowered, 2016) or Wander (Meta, 2019), a user can stroll the streets of Tokyo, climb Mount Everest, or see the pyramids in Egypt up close. Such solutions are safe, convenient, and educational, as well as accessible to people who, for various reasons, cannot travel.

Extreme experience simulators, such as virtual roller coasters, airplane flights, skydiving, or scuba diving, can not be ignored. Such applications merge excitement and adrenaline with safety, making it possible for users to experience something unique without risk.

VR does have its place in the adult entertainment industry – many companies are creating interactive experiences and 360-degree films that provide immersion and a sense of presence. This is a very controversial but fast-growing market segment that represents a large proportion of commercial VR use. This area also raises important ethical, psychological, and privacy concerns that are increasingly discussed in academia and the technology industry.

VR could drastically change how people consume entertainment and culture. It allows for active virtual participation, interaction, and experiences that were previously impossible in this form. Games, movies, concerts, and other cultural events in VR are offering something beyond the media themselves: emotion, movement, engagement, and the sense of presence. In the future, it surely will be even more realistic and ubiquitous, fusing the physical and digital worlds into a single, shared experience.

2.2 AUGMENTED REALITY IN INTERACTIVE ENTERTAINMENT

In the entertainment field, augmented reality (AR) technology also opens new possibilities. AR let players interact with the real world while at the same time virtual elements are added/seen via smartphone, tablet, or AR goggles. This allows users to not only observe a screen but also become part of the game or an event in their everyday environment. Players can discover hidden objects in the near place, take part in interactive challenges, or fight or compete with virtual characters who 'appear' in their world.

An example of AR game is Pokémon Go (Bueno et al., 2020). This is a mobile game in which users have to 'catch' virtual Pokémons. These creatures are visible in the real world but when looking through the phone's camera. The game encourages user to get out of their house instead of sitting in front of the screen. It also let users walk outside, explore, and engage in physical activity. This entertainment model allows players to move around in a real-world environment rather than sitting idly in front of a screen, which ultimately fosters activity and social integration. Another example is Harry Potter: Wizards Unite (Fandom, 2021) – a game in which users wander the streets, parks, or city centres and, using smartphones or AR goggles, participate in an adventure connected to a magical world, casting spells, discovering artifacts, and encountering magical creatures. However, it must be added that of today the game was shut down in 2022. Despite this, thanks to such titles, AR extends entertainment beyond four walls and allows the user to actively participate in the experience. Games like Pokémon Go have a very positive impact: they are changing the paradigm of typical gaming, which until now largely involved players locking themselves in a room in front of a computer or console and playing while seated. Traditional playing game results in prolonged sitting, poor posture of the body, weakened spines, eye strain, and limited movement. It often, but not only, refers to young people who can play games for a long time. In AR games, users stand, walk, and move. This can be good for physical posture and relationships with their surroundings. For example, a study showed that Pokémon Go players took significantly more steps on average in their first week of playing this game than before they started playing it (Philippe et al., 2024). However, it is worth noting that playing such games outside can also bring some risks. This, in turn can lead to accidents, getting lost, and reduced awareness. A paper (Wagner-Greene et al., 2017) examined the statistical aspects of game usage. Authors noted an increase in road accidents, collisions, and other incidents due to playing Pokémon Go game. Research proves a significant increase in the risk of incidents near locations where players play Pokémon Go. Numerous publicized incidents were also observed. What is more, because of playing this game, distracted drivers were also involved in accidents. Therefore, safety, proper control, and user awareness are so crucial. Figure 2.1 shows an example of how the AR module works in a smartphone application that searches for creatures outdoors.

AR in the sports and physical activity gaming segment is developing rapidly. An example is HADO which is an AR e-sports game. It was created by Japanese studio Meleap Inc. In this game, players wear special AR goggles and an sensors on arm(s). They play in teams throwing virtual

FIGURE 2.1 An illustrative example of how the AR module works in a smartphone application that searches for creatures outdoors.

energy balls and setting up virtual targets. This takes place in a sports hall or dedicated area (Marr, 2024). The game is a combination of fun, competition, and physical activity. Players jump, dodge, grab, and must avoid attacks. Such a game engage them and is an interesting experience for younger, as well as, adult users.

When it comes to entertainment and health perspective, AR sports games offer many benefits. Users are active in the game not only through using controllers and clicking but also because they move, use their coordination, reflexes, and physical strength. Scientific research proves that AR games can contribute to increased physical activity, improved motor coordination, and reduced sedentary behaviour (Philippe et al., 2024). In other words, AR active games can be good for people's health. In practice, AR can be used in promoting health. This can refer especially to children and adolescents. AR games can be interesting for them. They combine fun and movement and at the same time, they can be more encouraging and effective than traditional sport.

It is also worth mentioning AR applications in the context of multiplayer events, e-sports leagues, and public events. For example, there are AR arenas and gaming zones in entertainment centres. They offer team competitions using AR. In this way, AR can be described as not only individual gaming but also social one.

One interesting segment of AR entertainment is music apps and virtual instruments created using AR. Users can play a virtual piano, guitar, or drum set – ergonomically placed in their surroundings, visible through a smartphone, tablet, or AR goggles. For example, the 'Guitar 3D AR' (application for iOS) allows users to learn chords and compose music. They observe an animated visualization of the fingers moving over the instrument's socket. Such a solution combine entertainment with education. Players not only have fun but also develop musical skills. What is more, they can work on their coordination, concentration, and even move with gestures. These applications support active participation – the user must react, use their hands, and engage their body, distinguishing them from passively watching a video or playing on a computer. Moreover, institutions and entertainment centres are increasingly installing AR music stations in shopping malls and public spaces – where passersby can put on AR glasses or use a tablet, while a virtual piano or drums appear in the foreground, making them part of the performance. This format attracts attention, engages, and acts as an attractive form of promotion and interaction. Figure 2.2 shows an illustrative example of playing an instrument in AR.

FIGURE 2.2 Illustrative example of playing an instrument in AR.

AR is also entering the world of live culture and entertainment – concerts, exhibitions, performances, and outdoor events. AR installations are increasingly common in shopping malls, galleries, and festivals – interactive mirrors, displays that respond to user movement, and projections that merge reality with digital effects. For example, the National Geographic and INDE created the 'Air, Land & Sea' project. This project is an AR experience. This one allows users to virtually visit unique natural sites. It shows a dynamic, 'watering hole' and various animal species around it in their natural world. As part of the experience, viewers can observe animated animal models representing fauna from Africa to the Arctic, listen to rich background audio, and interact with presentation elements, providing both educational value and a high degree of immersion. The INDE BroadcastAR system, which underpins this installation, allows for the display of personalized, interactive 3D content on any large screen – from digital media inside galleries and museums to outdoor LED panels. The solution is flexible in terms of installation and is suitable for event, retail, and educational spaces. Additionally, the SnapShare module allows for the quick creation of photos and recordings of the experience and their sharing on social media or printing on-site, which increases audience engagement and provides a compelling promotional and branding opportunity. This AR mechanism places people in the physical spaces and enriches the cultural experience. This type of entertainment has a hybrid form. It merges the real world with the digital one.

AR in entertainment and live performance changes the way people engage with games, music, culture, and public spaces. It provides active participation and interaction. It brings something new which was previously impossible to experience. Mobile and desktop games, sports and educational AR apps, music, virtual instruments, cultural events, and interactive installations are just some of what's already available – what will continue to evolve. AR blurs the lines between 'online' and 'offline', and the user becomes an active participant, not just a spectator. Such experiences engage the body, mind, and emotions. Naturally, they pose new challenges (such as security, privacy, spatial control), but they contribute to a new lifestyle. This lifestyle is based on movement and the fusion of reality and technology.

While AR has many positive effects, such as developing imagination, encouraging movement, and allowing for new forms of interaction with the world, it is not without risks. One problem is the blurring of the line between the real and virtual worlds, especially for children and adolescents. People who frequently use AR games or applications may have difficulty distinguishing between what is real and what is merely a digital illusion. In young children, this can lead to disorientation and distorted perception of reality. Furthermore, many AR games feature the theme of fighting or eliminating opponents – virtual monsters or creatures. Repeated experiences of this type can, in some cases, reduce players' emotional sensitivity and even lead to increased aggression or a lack of empathy towards animals or other people in the real world. Some researchers notice that frequent exposure to digital violence, even if it is purely for fun, can weaken the boundary between fiction and reality. Such experiences can be also addictive. When players play realistic and immersive games, it can be difficult for them to return to a peaceful, everyday life. On the one hand, technologies such as AR should be developed. But on the other hand, education in this area should be taken care of. Furthermore, the responsible content provided by AR should also be taken care of. It can help to prevent entertainment from becoming a source of dangers. If this is not ensured, the use of AR can have very negative consequences.

Another risk of AR is physical safety. During AR games, like 'Pokémon Go', or other applications when navigation is required, users often focus their eyes on either the smartphone or headset screen, not on what happens around them. This may lead to certain bad consequences: accidents, falls, walking into the street without looking, and even collisions with other people or different kinds of obstacles. There have been cases where players, while being deeply involved in the search for virtual objects, have entered private property, dangerous places, or spontaneously appeared near busy roads. Children are at especial risk because they do not always have adequate awareness of hazards in the real world. Another problem is distraction. Using AR while driving or cycling carries a real risk of accidents. That is why application and device manufacturers increasingly add warning messages and systems limiting the work of AR in potentially dangerous places. However, the greatest responsibility lies with users themselves and their guardians, who should not forget that even the most realistic game is not capable of taking the place of common sense and real-life caution.

2.3 IMMERSION EXPERIENCES IN VIRTUAL AND AUGMENTED REALITY

VR and AR are often described as immersive. It means that they immerse users in an alternative or enhanced version of reality. Immersion in this context refers to the depth of engagement that they experience when interacting with a digital environment. This engagement is on different levels: mental, emotional, and sensory. In VR, users are fully surrounded by a virtual world. In AR, digital elements display on their real surroundings. In both cases, the line between what is real and what is artificial blurs. This gives a sense of 'being in' another world. After a long time of using VR or AR, the player may become confused.

One of the most characteristic aspects of immersive technologies is their ability to manipulate human perception. When a user puts on a VR/AR headset, their brain quickly adapts to the simulated environment. Visual, auditory, and sometimes tactile stimuli enhance the illusion of presence.

Then mind can perceive virtual elements as real. Therefore, people can instinctively reach for a digital object and move away from a virtual obstacle. They can even experience emotions such as fear, excitement, or awe. These emotions are normally reserved for the real world. However, they can be stimulated via virtual world too. This often happens: users know they are in a real room. They also know there is a wall nearby. Yet, when they see an object in the virtual world, they try to touch it. Often, they hit the real wall with their hand or controller. This demonstrates just how immersive the virtual world can be.

Some main features of immersive VR and AR entertainment are:

- Realistic Presence: Users feel as if they are really inside the virtual or augmented world, even though they are not.
- Sensory Immersion: It uses sight, sound, and sometimes touch or movement to make the experience more real.
- Interactivity: Users can move around, touch, and change things in the virtual world. This makes them more involved.
- Mix of Real and Digital Worlds: In AR, virtual objects appear in the real world and react to what users do.
- Emotional Impact: The sounds and images can make users feel emotions similar to real-life situations.
- Loss of Time and Place: Users often forget where they are and focus only on what they see in VR.
- Freedom from Limits: VR and AR let people do things that are impossible in real life, like flying or visiting distant places.

This strong level of engagement has both good and bad sides. On the positive side, immersion lets people live through amazing situations – like flying in space, diving in the ocean, exploring ancient places, or playing instruments that don't really exist. It can also help build empathy, support creativity, and improve memory, especially in learning or art. However, too much immersion can make people lose touch with reality. Users may forget about time, feel confused when they return to the real world, or even mix up what was real and what was virtual. Research by Taveira et al. (Taveira et al., 2022) shows that very intense virtual experiences can cause short-term feelings of detachment from oneself or the world. The effect usually doesn't last long, but for some users, especially those more sensitive, it can be uncomfortable. Another study by Girondini et al. (Girondini et al., 2024) found that after spending a lot of time in VR, users sometimes have trouble judging the size or distance of real objects. This means that deep VR sessions can temporarily change how the brain sees and measures space.

Immersive experiences bring excitement and strong emotions, but they can also cause stress or even addiction. The stronger the feeling of being 'inside' the virtual world, the easier it is to forget about the real one. This deep immersion can create a trance-like state where users feel fully absorbed and emotionally connected to what they see and hear. Sometimes it can be inspiring and relaxing, but for others, it may be confusing or addictive. Because of this, both creators and users should remember that immersion is more than just fun – it can deeply affect how people think, feel, and behave even after they take off the headset.

Spending too much time in perfect virtual worlds can lower real-life happiness. When people see places that look better than real life, the real world may start to feel boring or disappointing. This can cause sadness, low mood, or what some call 'post-immersion blues'. Also, watching too many ideal or fake images, for example, in virtual relationships, can change how people see others. It may lead to frustration, loneliness, or problems with real emotional connections. Table 2.2 presents summary of advantages and disadvantages of immersion in VR and AR technologies.

Immersion in virtual and augmented reality technologies is one of the most important phenomena in contemporary digital culture. It allows users to interact with new worlds, experience emotions, and discover unknown places. Users can develop their senses in ways that were previously impossible. At the same time, such deep immersion in the digital world is connected with many risks. They can have two forms, both psychological and physical. Immersion in VR can blur the lines between the real and virtual worlds. It can even affect how users perceive themselves and their surroundings. Therefore, it is important to link the development of immersive technologies with ethical reflection, education, and responsible content design. Properly utilized, VR and AR can become tools not only for entertainment but also for development, education, and therapy – provided that a balance between the virtual and real worlds is maintained.

TABLE 2.2 Advantages and Disadvantages of Immersion in VR and AR Technologies

Aspect	Advantages	Disadvantages
User Experience	• Deep emotional and cognitive engagement. • A sense of 'presence' in the virtual world. • Possibility of full interaction and control over the environment.	• Risk of losing orientation in reality. • Blurring of the line between the real and digital worlds. • A feeling of disorientation after the session ends.
Psychological Aspects	• Development of empathy (e.g., by experiencing another's perspective). • Increased creativity and emotional expression. • Potential therapeutic applications (e.g., in VR therapy for anxiety).	• Possible symptoms of depersonalization or derealization. • Risk of addiction to virtual experiences. • Emotional overstimulation or stress.
Perception and Senses	• Stimulation of sight, hearing, and touch. • Enhanced perception of space and movement. • Realistic representation of stimuli.	• Sensory overload (e.g., dizziness, motion sickness). • Temporary disturbances in the perception of distance or proportion. • Visual and nervous system fatigue.
Educational and Cognitive Aspects	• Highly effective experiential learning. • Easier memorization, thanks to emotions and active participation. • Possibility to explore inaccessible areas.	• Risk of superficial knowledge acquisition, focusing on impressions instead of content. • Overdependence on technology.
Social Aspects	• New forms of collaboration and communication in the digital environment. • Shared experiences despite physical distance.	• Isolation from real-life contact with people. • Difficulty returning to the 'real' social world. • Reluctance to return to ordinary reality/relationships with real people after experiencing an idealized world; 'post-immersion sadness'. • Distorted expectations about the real world and people.

Source: Author's own work.

BIBLIOGRAPHY

Bueno, S., Gallego, M. D., & Noyes, J. (2020). Uses and gratifications on augmented reality games: An examination of Pokémon GO. *Applied Sciences, 10*(5), 1644.

Fandom. (2021). *Harry Potter: Wizards Unite*. https://harrypotter.fandom.com/pl/wiki/Harry_Potter:_Wizards_Unite

Girondini, M., Montanaro, M., & Gallace, A. (2024). Exploring virtual reality object perception following sensory-motor interactions with different visuo-haptic collider properties. *Scientific Reports, 14*(1), 10011. https://doi.org/10.1038/s41598-024-59570-x

Haptx. (2024). *Experience the power of touch in VR and robotics*. https://haptx.com/

Holodia. (2025). *Holofit VR Fitness Virtual Workouts, Real Results*. https://www.holodia.com/

KAT VR. (2025). *KAT Walk C 2 Core*. https://www.kat-vr.com/?variant=51488283394353

La Biennale di Venezia. (2023). *Venice VR Expanded*. https://www.labiennale.org/en/cinema/2021/lineup/venice-vr-expanded

Marr, B. (2024). *HADO: The Amazing Ways Augmented Reality Is Changing Sport*. https://bernardmarr.com/hado-the-amazing-ways-augmented-reality-is-changing-sport/

Meta. (2019). *Wander*. https://www.meta.com/pl-pl/experiences/wander/2078376005587859/?srsltid=AfmBOorZ8IxYaT0zjLcOJHzvXHmz-CsXCK2cY7KLKCFWAwq-3J3CRJGK

Philippe, A. G., Goncalves, A., Korchi, K., & Deshayes, M. (2024). Exergaming in augmented reality is tailor-made for aerobic training and enjoyment among healthy young adults. *Frontiers in Public Health, 12*, 1307382.

Senseglove. (2025). *Training the workforce of the future*. https://www.senseglove.com/

Steampowered. (2015). *Keep Talking and Nobody Explodes*. https://store.steampowered.com/app/341800/Keep_Talking_and_Nobody_Explodes/

Steampowered. (2016). *Google Earth VR*. https://store.steampowered.com/app/348250/Google_Earth_VR/?l=polish

Stecuła, K. (2022). Virtual reality applications market analysis—On the example of steam digital platform. *Informatics, 9*(4), 100.

Stecuła, K. (2024). Analysis of asymmetric VR games—Steam platform case study. *Technology in Society, 78*, 102673. https://doi.org/10.1016/j.techsoc.2024.102673

Stecuła, K., & Naramski, M. (2025). Virtual reality as a green tourism alternative: Social acceptance and perception. *Sustainability, 17*(17), 7722.

Taveira, M. C., de Sá, J., & da Rosa, M. G. (2022). Virtual reality-induced dissociative symptoms: A retrospective study. *Games for Health Journal, 11*(4), 262–267.

Virtuix. (2025). *Omni One*. https://virtuix.com/

VIRZOOM. (2024). *VIRZOOM*. https://virzoom.com/

Wagner-Greene, V. R., Wotring, A. J., Castor, T., Kruger, J., Mortemore, S., & Dake, J. A. (2017). Pokémon GO: Healthy or harmful? *American Journal of Public Health, 107*(1), 35.

Wave. (2025). *Wave's mission is to unite humanity through immersive 3d entertainment*. https://wavexr.com/about/

CHAPTER 3

Shopping and Retail

3.1 APPLICATIONS FOR SHOPPING IN VIRTUAL REALITY

Virtual reality (VR) is becoming one of the most innovative tools in modern commerce. With the rapid growth of online shopping and the continuing popularity of retail, consumers are seeking new and comfortable ways of shopping. Traditional online stores let clients see the products in 2D. This limits the ability to fully assess product's appearance, size, and functionality. In this context, VR is a natural step forward. VR allows customers to fully immerse themselves in the digital store environment. Clients can view products in 3D, and even interact with them. For businesses, VR offers an opportunity to build a competitive advantage and increase customer engagement. VR let offer customers modern and convenient way of shopping. Below, examples of VR applications in shopping and retail are presented.

Solutions already exist that allow customers to visit stores using VR. They can enter a digital store from their home. All they need is VR headset and controllers. Sometimes, they do not even need controllers – hand movements are sufficient. Customers can navigate virtual aisles within a given store, and even selecting and viewing products (Stecuła et al., 2024). VR gives a sense of being in a physical store. Thanks to this solution, a client does not need to visit a shop physically. This saves some time and effort. It makes shopping more convenient and more tailored to customer needs. It is worth noting that online shopping via a website or application is not as immersive as in VR. Figure 3.1 shows an illustrative drawing of what it looks like to browse VR products.

DOI: 10.1201/9781003643395-3

FIGURE 3.1 An illustrative drawing of VR shopping.

Such solutions can be used in many industries, the first of which is clothing retail sector. In this industry, VR technology opens up entirely new shopping possibilities that go beyond the traditional store and website. Virtual fitting rooms are equipped with VR headsets and allow customers to immerse in a digital showroom. There, customers can rotate around clothing models, change colours, styles, and even check the fit in a three-dimensional space. Such solutions offer the advantages of online and in-store shopping. On the one hand, customer can experience immersion, and on the other – interact with products. A study by Ricci et al. (2023) shows that the use of immersive VR environments in fashion increases user engagement and improves their satisfaction with the shopping experience. Interesting results can be found in Chen et al.'s (2024) literature review. Authors indicate that virtual try-on (VTO) technologies, which can operate in VR or augmented reality (AR) environments, significantly influence consumer purchase intentions. They refer to key factors such as usability, aesthetics, emotion, and ease of use. With VR, clothing stores can offer customers a more convenient and interactive shopping experience – reducing the perceived risk of online purchases, reducing returns, and building a competitive advantage through modern shopping experiences.

When it comes to the food industry, Tesco has developed 3D e-commerce offerings. In 2012, it created a 3D virtual store that clients can visit via smart TVs (Wagaine, 2022). Clients can use motion sensor controls to walk in the virtual store. In this store, they can watch and buy products. On the other hand, Tesco in South Korea developed innovative virtual shops. Such virtual stores were available in metro stations and bus stops. They display products on walls (Businesstoday, 2015). It work as follows: clients need to scan the QR code of the selected product using the Tesco Homeplus application downloaded on smartphone. Then, the product is delivered directly to the client's home. It is also possible to choose a delivery time (Archello, 2012). Such applications help people, especially busy ones, to place grocery orders while waiting for trains or buses. It saves time and is an effective way of doing the convenient shopping. Such a solution creates a new market adjusted to the country's demanding work culture.

Amazon introduced VR kiosks in shopping malls (Emory, 2020). During the 'Prime Day', customers had a chance to transfer themselves to the special city full of Prime Day products. It was possible by using VR goggles (Oculus Rift) and controllers (Oculus Touch). Clients interacted with products in virtual 3D environment. They could see, touch, and watch different type of products. Such a solution combines the convenience of online shopping with the sensory engagement of brick-and-mortar retail.

HappyFresh developed a service called HappyReality. It is the first and fastest-growing online grocery platform in Southeast Asia (HappyFresh, 2015). The solution let clients experience VR grocery shopping (Yahoo!, 2017). The quality of shopping is quite high due to the realistic images, sounds, and other sensations from a real-world. Additionally, client can activate the special function to identify their favorite brands. It makes shopping adjusted to the very client and personalized. This service helps customers to make a decision and to find the most popular products in any category. According to some authors (Ketoma et al., 2018), VR grocery shopping solves some of the limitations of online grocery shopping. But there are still some challenges connected with delivery, product quality, and service quality. It is worth noting that virtual supermarkets in the form of 360-degree video have a chance to create highly immersive sensory experiences. Such supermarkets encourage consumers to stay in the store, affect their senses, touch their hearts, and stimulate their minds (Violante et al., 2019).

In the furniture and interior design industry, VR is also used. Customers can 'enter' a virtual showroom or room where they arrange furniture, test its layout, colours, and functionality in 3D. This allows them to better visualize what a given sofa, table, or shelf will look like. This reduces the risk of a bad purchase and reduces the number of returns. Research by Lee et al. (2024) shows how creating fully functional VR stores – including those in the furniture segment – impacts the quality of the user experience and their confidence in shopping in a virtual environment. For businesses in this industry, VR can be a competitive strategy tool. It allows them to show their entire offerings in an environment that does not require a physical showroom. This, in turn, can reduce costs and expand reach. From the customer's perspective, the experience becomes more engaging and comfortable. Instead of scrolling through online photos, they can 'walk' through the virtual space. It let them try out different options. This experience can increase purchase satisfaction and build stronger brand loyalty.

The accessories, watches, and jewellery industries are also using VR to create more engaging shopping experiences. Customers can look at luxury watches and jewellery in a virtual space. They can turn the items, zoom in on small details, and see them in different light or on various avatars. Chang and Lekena (2024) studied how VTO – works for luxury goods. They found that this technology makes buyers feel more sure about their choice. VR helps brands show their products in elegant, digital showrooms. It removes the need for real stores in every city. People can enjoy an experience similar to a real visit, but online. This is especially important for young and global customers. Literature suggests that VR systems improve product visualization and can influence purchase intentions in e-commerce (Liu et al., 2020).

In the consumer electronics and automotive sectors, VR technologies are being used to present products in ways that exceed the capabilities of traditional catalogs. Customers can 'enter' virtual showrooms, view new equipment (TVs, speakers, home appliances) or full-scale cars, and change configurations, colours, interiors, and the user environment. In the automotive retail context, VR allows for virtual test drives and vehicle interior exploration without physically being present in the showroom. This can help increase customer engagement and shorten the purchasing process. Companies utilizing such technologies can gain a competitive advantage by offering a superior shopping experience tailored to the needs of digital customers. In the paper of Mekala and Arun Ganessh (2025), it is analyzed how virtual car showrooms and VR test drives influence purchasing decisions and customer experiences. The results show that VR can streamline the car buying process, allowing customers to safely and realistically experience vehicle features and personalize their experience. It's also worth noting that the lack of tactile feedback and the potential for motion sickness are limitations. Differences in technology acceptance between demographic groups were also noted, with younger users demonstrating greater enthusiasm for such solutions. The paper indicates that VR can effectively complement traditional showrooms, increasing customer engagement and brand trust.

Shopping in VR may also be related to the concept of the metaverse. The concept of the metaverse can be defined as a collective, shared virtual space created by combining a virtually enhanced physical reality with physically persistent virtual spaces (Cheng, 2023; Mystakidis, 2022). It has a potential in online shopping. In metaverse, users (clients) can take part in interactive shopping experience that combine the convenience of online shopping with the sensory and social aspects of in-store shopping. When it comes to the shopping experience in metaverse, it is possible for users to (Stecuła et al., 2024):

- Navigate virtual stores using avatars.
- Interact with 3D models of food products.
- Receive real-time assistance from virtual salespeople.
- View interactive product presentations.
- Access personalized recommendations.
- Use advanced search options.
- Create your own shopping lists.
- Interact with other people in virtual stores.

The store in metaverse can virtually reproduce a real store. There, clients can walk around shelves, watch and check products, meet other people, be advised by virtual assistants, and have a preview of their product cart. As technology develops, the metaverse can meaningfully change the shopping and retail industry. It can provide a shopping experience that addresses the gap between the physical and digital worlds.

Table 3.1 summarizes the usage of VR in shopping and retail.

TABLE 3.1 The Usage of Virtual Reality in Shopping and Retail

Industry	Examples of VR Usage
Fashion and Clothing	• Virtual fitting rooms. • 3D try-on of clothes. • Interaction with 3D products. • Change of colours, styles, and sizes.
Food and Grocery	• Virtual supermarkets. • 3D shelves and product displays. • QR code product selection. • Online delivery from VR store. • Virtual shopping while waiting (e.g., metro or bus stops).
Furniture and Interior Design	• 3D visualization of rooms. • Virtual arrangement of furniture. • Testing colours and layouts.
Accessories, Watches, Jewellery	• 3D product viewing. • Virtual try-on for watches and jewellery. • Viewing under different lighting. • Luxury virtual showrooms.
Electronics and Automotive	• Virtual showrooms. • 3D exploration of cars and devices. • Virtual test drives.

Source: Author's own work.

3.2 AUGMENTED REALITY SOLUTIONS IN RETAIL

The way people do the shopping is rapidly changing, influenced by the development of new technologies. ARis one of the most innovative tools changing the retail world. It helps connect real shopping with digital experiences. With AR, customers can see and explore products in new ways. They can check details, compare options, and make better buying choices. The technology adds digital elements to the real world, giving more information while shopping. People can use their phones or AR glasses to find products, read details, or see how items look at home before buying. These tools make shopping more personal and fun. They also help brands build stronger relationships with customers and increase loyalty (Wolniak et al., 2024).

The first example of using AR in shopping is augmented packaging. They allow brands to provide additional product information by scanning a QR code or graphic on a label (Wagaine, 2022). An example is the Italian sauce brand Francesco Rinaldi, whose products, after scanning a phone, play animations and content about the product's ingredients and origin (Spinner, 2018). This technology not only provides information but also strengthens the brand's image. It can be done through storytelling elements

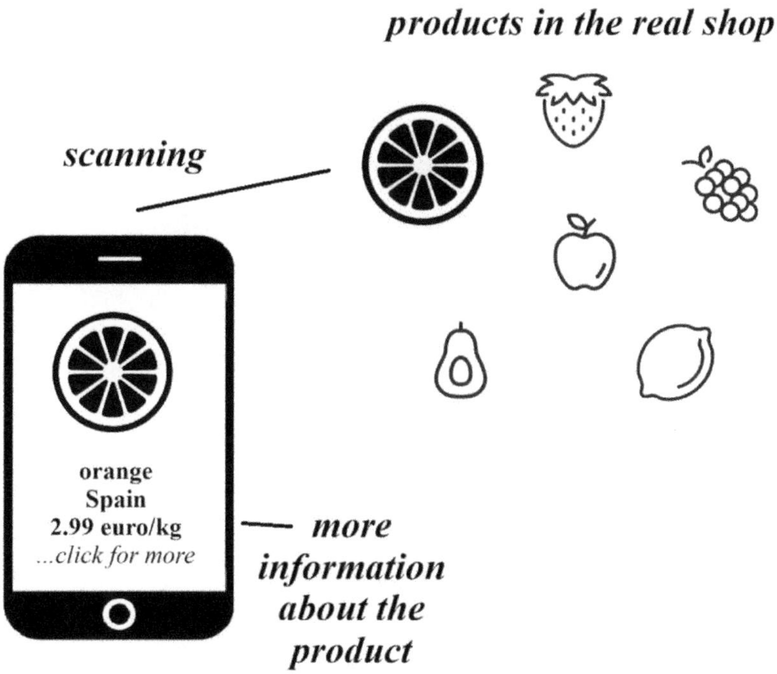

FIGURE 3.2 Illustration of the use of AR in a store – scanning products with a smartphone to see additional product information.

and engaging consumer experiences. Jack Daniel's WebAR allows users to virtually visit a distillery in Lynchburg in Tennessee (Wagaine, 2022). Another article presents a web prototype that allows smartphones, tablets, and AR glasses to display product data in stores (Xu et al., 2022). This helps to identify the product and see detailed information (for example, product descriptions or reviews). This element is then displayed directly on the products that customer wants to buy. This solution aims to help consumers make purchasing decisions, shorten decision-making time, and facilitate product comparison. Figure 3.2 presents an illustrative drawing showing the technology of scanning products in a store with a smartphone in order to obtain more information about the product.

Another example of the practical application of AR in grocery retail is the List&Go application. It was developed by Marks & Spencer (M&S) and tested in the UK (Nott, 2022). Thanks to this application, customers can see a shopping list and then, be guided through the store to the very product. They can see a path (virtual path) to specific products on the shelves via their smartphones. The system utilizes Wi-Fi and digital

FIGURE 3.3 Illustrative drawing of the use of augmented reality navigation in a store to guide the user to a selected product (milk carton).

store planograms, marking the first implementation of its kind in the UK. Figure 3.3 shows an illustration of finding a path to the milk carton.

The National Independent Supermarkets Association (NISA) retail chain worked with the company Jisp to create the Scan&Save system. It uses AR to make shopping easier. Customers scan product barcodes to see special offers and use coupons right on their phones (Jisp Bussiness Solutions, 2022). During the pilot program, the results were very good – more than 82,000 scans, 40,000 clicks, and 32,000 coupons redeemed. As many as 81% of users used the app repeatedly. Additional benefits of the implementation include reduced paper consumption and reduced food waste.

From a scientific research perspective, AR technology in grocery retail is increasingly finding its effectiveness. In some research, the authors demonstrated that AR smartphone apps can effectively support customers in making healthier food choices (Ahn et al., 2015). Their application offered personalized recommendations for healthy products and warned against unsuitable choices based on the user's individual needs (e.g., allergies or a low-sodium diet). User studies have shown that AR overlays significantly reduce product search time and improve in-store orientation. Similar research was conducted by other authors, who developed the mobile assistant PHARA to support food choices in grocery stores (Gutiérrez et al., 2018). The study confirmed that properly designed AR interfaces allow customers to interpret product information more quickly and effectively. Another AR mobile application also helps customers make healthier choices in grocery stores (Ahn et al., 2015). The system suggests products to avoid (e.g., due to allergies or a low-sodium diet) and recommends products. Using the app has been scientifically tested, and AR labeling has been shown to reduce the time it takes to find healthy options.

Other researchers have developed the ARGrocery application, which uses marker-based AR to identify products using a smartphone camera (Yew, 2023). The application presents data on the product's name, brand, and ingredients using coloured AR markers that help quickly distinguish recommended products from those not recommended.

Another solution is CoShopper, a system that combines artificial intelligence and AR to streamline grocery shopping (Alhamdan et al., 2020). It uses CNN neural networks to recognize products. It is also based on data from nutrition databases, medical reports, and stores to provide nutritional information in current time.

In the literature, there is also a paper that presents a system combining deep learning with AR. The aim of this is to improve customer navigation in stores and help them get information on the product (Cruz et al., 2019). The system recognizes the store layout with 98% accuracy, indicates the customer's location and route to the product, and also enables 3D visualization and analysis of user behaviour.

Another practical solution is the system called the Third-Eye. It was created mainly for people who are blind or have low vision (Advani et al., 2016). The system uses AR glasses, a glove with sensors, and a smart shopping cart. These tools help users find products, move around the store, and pick groceries.

AR also has great potential in the clothing and fashion industries. More and more of this sector is going digital. As online shopping grows, people want shopping to feel more real and interactive. Stores compete by creating fun and easy digital tools for customers. AR fits this trend well. It lets people try on clothes, shoes, or accessories without going to a store. With a phone camera or AR mirror, they can see themselves in different outfits, check sizes, colours, and styles, and find matching pieces. This works both at home and in physical stores. AR mirrors are one example (Manocha et al., 2024; Xue, 2022). A customer looks into the mirror, picks clothes, and sees how they look instantly. By looking at themselves in the mirror's reflection, they can see how they look in the outfit. Such solutions can increase customer satisfaction, reduce returns, and enable fashion brands to build a modern, interactive image. Figure 3.4 shows illustrative drawing of how AR mirror works.

In the apparel and footwear industry, AR allows customers to virtually try on clothes, shoes, and accessories without the need for physical try-on. VTO applications allow the use of a user's photo or avatar, onto which clothes are 'overlaid' in 2D or 3D, often with material deformation effects to make them appear natural with movement. When it comes to the literature, some paper (Chen et al., 2024) analyse 69 articles in this topic. It concludes that factors such as perceived utility value, aesthetics, emotions, and ease of use have significant impact on consumer buying decisions using VTO.

FIGURE 3.4 An illustrative drawing of how AR mirror works.

Another example describes FITMI's solution. This is a VTO model that improves the realism of online fitting rooms (Samy et al., 2025). The proposed system helps connect traditional store try-ons with online shopping. FITMI gives users an interactive virtual try-on experience. It uses latent diffusion models to create high-quality images with clear textures. As a web application, FITMI lets users upload their photos and try on clothes from online catalogs. This gives a realistic view of how the items would look in real life. FITMI also works as a style assistant. It recommends matching products, shows similar options, and suggests outfits based on user preferences. This makes shopping more enjoyable and helps reduce returns caused by poor fit or wrong choices.

Researchers also describe an AR mobile application that uses 3D modeling to show realistic shoe models (Shrestha et al., 2025). Users can see how a shoe looks in their space through AR – like a virtual shoe fitting. This is especially useful for fashion brands and e-commerce.

Another example is AR for eyewear and watches (Konarzewski & Reiner, 2023). Virtual try-on tools in these categories affect how millennial shop in the DACH region. Such applications make buying decisions easier and improve the overall shopping experience. Consequently, individuals readily switch from in-store shopping to browsing and purchasing eyewear online. However, it's important to emphasize that while this solution may be beneficial for large eyewear retailers, smaller retailers struggle with the significant capital expenditure associated with implementing this technology.

Other articles also present solutions for VTO. Other authors discuss trying on high-end accessories, including watches and jewellery (Chang & Lekena, 2024). Generative models and hand tracking techniques are adapted to enable realistic watch trying-on in an AR environment.

Some authors (Ghatge et al., 2024) write about mobile AR solutions for virtual jewellery fitting. These are designed to enhance e-commerce user experiences by allowing them to visualize and interact with jewellery in real time. The system detects poses in and tracks body moves, for example, moves of arms and fingers. The app ensures precise application of virtual jewellery to detected positions. Interactive gesture-based controls, including touching, scaling, and rotating, allow users to customize the fit and appearance of jewellery. The study identifies new opportunities to increase customer engagement and drive e-commerce growth in the jewellery sector. In the literature, there are more practical examples of the usage of AR in jewellery, for example, the paper of Zhang-Wu (Zhang-Wu, 2024).

According to the author, AR replaces the initial personalization and try-on process. It also encourages interactive engagement. A realistic try-on experience helps customers make good buying decisions.

AR solutions can also apply to cosmetics. The article of An et al. (2023) presents the ARCosmetics system which enable try virtual cosmetics on using smartphones. The user can try on lipsticks, eyeshadows, and blushes, without the need to physically test them in a store. Conducted research has confirmed the system's effectiveness and high user satisfaction. Sephora also allows its clients to personalize their shopping experience by trying on makeup virtually using AR (Rayome, 2018).

AR can also be used in the furniture and home furnishings sector. It can have an impact on the way consumers do the shopping and plan their living spaces. The first example of application is IKEAPlace. It allows users to visualize how a given furniture (for example, sofa, coffee table, or chair) can look in their own room. Customers can check product's size, style, colour, and layout relative to other room elements (IKEA, 2017). In some articles (Ozturkcan, 2021), the author analyse how such applications allow a furniture company to change the way of shopping. It can reduce the risk of purchase errors and increase customer confidence. Another study shows that implementing AR in furniture retailing can increase sales and reduce returns. This is due to allowing customers to better match purchased products to their home environment (Gutierrez-Antunez & Leal-Enríquez, 2024). Moreover, AR is also used in consumer electronics and smart home equipment. Customers can see how a TV, speaker, or household appliance look like in their room. They can check if the product blends with the surrounding and whether its size and style are appropriate. In this way, AR serves as a bridge between online shopping and the physical experience. It reduces the risk of incorrect or inappropriate purchases, increases the comfort of purchasing decisions, and transforms the traditional process of selecting and ordering interior design elements.

As can be concluded from the examples discussed, AR has great potential in the field of shopping and retail. Table 3.2 summarizes the discussed use of AR in shopping and retail.

The development of AR applications in retail shows how this technology changes shopping. Summing up, AR can navigate and guide customers through stores but also can offer interactive coupons and packaging. All the mentioned solutions increase consumer engagement, improve orientation in the store, and help in building a brand. As AR technology

TABLE 3.2 The Usage of Augmented Reality in Shopping and Retail

Industry	Examples of AR Usage
Grocery Retail	• Scanning products to display additional information. • In-store navigation guiding users to selected products. • Displaying promotions and coupons after scanning products. • Supporting healthier choices through AR product labeling.
Clothing and Footwear	• Virtual try-on systems for clothes and shoes without physical fitting. • AR mirrors allowing users to see themselves in selected outfits. • Personalized styling and size or colour matching.
Accessories, Watches, and Jewellery	• Real-time virtual try-on for eyewear. • Virtual watch fitting using hand-tracking technology. • AR jewellery fitting (rings, necklaces, bracelets) with gesture-based interaction.
Cosmetics and Beauty	• Real-time virtual try-on for makeup products (lipsticks, eyeshadows, blush) on smartphones. • Personalizing makeup and selecting shades suitable for skin tone. • AR mirrors in beauty stores for interactive testing.
Furniture, Electronics, and Home Décor	• Visualizing furniture in real space before purchase. • Checking colour, size, and style fit within the room. • Viewing 3D furniture models in true scale. • Previewing placement of home appliances and décor elements. • Visualizing electronic devices (TVs, speakers, home appliances) in the user's room. • Integrating AR with smart home systems to preview device functions. • Assisting in room layout planning and device placement.

Source: Author's own work.

continues to advance, retail is moving towards even greater personalization, convenience, and immersion, potentially transforming the way consumers shop in the future.

3.3 MARKETING IN THE CONTEXT OF VIRTUAL AND AUGMENTED REALITY

VR gives retail and e-commerce brands new ways to shape the shopping experience. With VR, customers can enter virtual stores or showrooms and move through 3D spaces. They can look at products, try them on, test features, and make purchases – all in a simulated environment. This changes traditional marketing methods. Instead of just browsing a website or visiting a store, shoppers take part in a full, sensory experience. Research shows that customers who explored a store in VR felt more positive emotions and rated the store as more attractive than those using a regular website (Jin et al., 2021). Henriques and Winkler (Henriques & Winkler, 2021) studied how VR is used in the automotive industry, especially in car clinics: early design stages for new models. They identified 16 key challenges and opportunities for VR use, such as cost, realism, depth and motion perception, product interaction, graphic quality, ease of use, haptic feedback, and simulator sickness. The authors note that VR has strong potential for marketing research and car design but still needs better hardware, software, and visual quality to create a convincing user experience.

In the marketing context, VR can be used to:

- Create virtual showrooms where products are presented on a realistic scale and with the possibility of interaction.
- Organize virtual promotional events (e.g., product launches, virtual test drives, virtual fashion shows).
- Personalize the shopping journey through customer avatars, analyse their behaviour in the VR environment, and customize recommendations in real time.

Marketers can thus build a brand that stands out from standard online and offline channels. From a business efficiency perspective, VR marketing also offers the opportunity to shorten time-to-market, test configurations, exposure, and customer behaviour in a simulation environment. However, VR marketing also poses challenges. They include the high cost of producing VR environment, the need for special equipment, and the risk of simulator sickness. There is also the need to ensure good quality visual stimulation and interaction to make the experience believable and engaging.

AR, in turn, allows marketers to add digital layers to the traditional products and shopping. For example, this can involve the overlay of 3D graphics, contextual information, interactive fitting rooms, as well as product visualizations, for example, in the customer's house. In marketing, AR is becoming not only a visualization tool but also an environment in which to increase user engagement, shorten the purchase path, and build brand loyalty. Some studies (Tan et al., 2022) discovered that the use of AR in a retailer's mobile application was correlated with lower purchase uncertainty, higher propensity to purchase, and a greater share of sales through the online channel. It refers especially to more expensive products and lesser-known brands. AR can be also used for catalogs. Some authors (Fachrurozi & Wibowo, 2023) in their article describe a catalog application with AR technology that displays furniture products in real time. It facilitates customers' furniture selection by presenting a three-dimensional image of the product from various angles.

Key areas of AR application in retail marketing include:

- VTO: For example, applications that allow customers to view clothes, shoes, makeup, or furniture in their real surroundings.
- Interactive Communication and Storytelling: AR allows brands to create campaigns in which packaging or posters activate additional content in the app (animations, games, product presentations).
- Increasing Informational Value: Using AR, customers can view a product in 3D, checking its features, fit, colours, and context (e.g., a piece of furniture in a room, a kitchen appliance).

For marketers, this means that when using AR, the focus should not only be on the 'wow factor', but also on delivering value to the user through interactivity, personalization, and context. For example, a 'shoppable AR' campaign allows consumers to scan a product or QR code and make a purchase directly from the AR application – integrating the marketing experience into the transaction (Fischer, 2018).

Implementing AR in retail marketing also comes with its own challenges – you need to ensure compatibility with user devices (smartphones, tablets), ease of use of the app, protection of customer privacy and data, and avoid overcomplicating the experience, which may discourage less technologically advanced consumers.

BIBLIOGRAPHY

Advani, S., Zientara, P., Shukla, N., Okafor, I., Irick, K., Sampson, J., Datta, S., & Narayanan, V. (2016). A multitask grocery assist system for the visually impaired: Smart glasses, gloves, and shopping carts provide auditory and tactile feedback. *IEEE Consumer Electronics Magazine*, *6*(1), 73–81.

Ahn, J., Williamson, J., Gartrell, M., Han, R., Lv, Q., & Mishra, S. (2015). Supporting healthy grocery shopping via mobile augmented reality. *ACM Transactions on Multimedia Computing, Communications, and Applications (TOMM)*, *12*(1s), 1–24.

Alhamdan, Y., Alabachi, S., & Khan, N. (2020). CoShopper-leveraging artificial intelligence for an enhanced augmented reality grocery shopping experience. *2020 IEEE International Conference on Artificial Intelligence and Virtual Reality (AIVR)*, 337–338.

An, S., Chen, J., Zhu, Z., Zhou, F., Yang, Y., Ma, Y., Liu, X., & Zhu, H. (2023). ARCosmetics: A real-time augmented reality cosmetics try-on system. *Frontiers of Computer Science*, *17*(4), 174706.

Archello. (2012). *Tesco opens worlds first virtual store.* https://archello.com/project/tesco-opens-worlds-first-virtual-store

Businesstoday. (2015). *How Tesco virtually created a new market on a country's lifestyle.* https://www.businesstoday.in/magazine/lbs-case-study/story/case-study-tesco-virtually-created-new-market-based-on-country-lifestyle-143807-2015-02-06?onetap=true

Chang, T.-Y., & Lekena, S. K. (2024). GlamTry: Advancing Virtual Try-On for High-End Accessories. *ArXiv Preprint ArXiv:2409.14553.* https://arxiv.org/abs/2409.14553

Chen, C., Ni, J., & Zhang, P. (2024). Virtual try-on systems in fashion consumption: A systematic review. *Applied Sciences*, *14*(24), 11839.

Cheng, S. (2023). Metaverse. In: *Metaverse: Concept, content and context* (pp. 1–23). Springer.

Cruz, E., Orts-Escolano, S., Gomez-Donoso, F., Rizo, C., Rangel, J. C., Mora, H., & Cazorla, M. (2019). An augmented reality application for improving shopping experience in large retail stores. *Virtual Reality*, *23*, 281–291.

Emory, C. (2020). *Will the Amazon VR kiosks reinvent retail shopping?* https://www.digitalbodies.net/will-the-amazon-vr-kiosks-reinvent-retail-shopping/

Fachrurozi, H., & Wibowo, A. P. (2023). Application of augmented reality for furniture catalogs. *Journal of Social Research*, *2*(11), 4040–4052.

Fischer, S. (2018). *Snapchat will allow users to buy products via augmented reality.* https://www.axios.com/2018/04/19/snapchat-brands-buy-shoppable-ar-augmented-reality

Ghatge, D., Deshmukh, S., Gaikwad, A., Iparkar, K., & Mane, A. (2024). Virtual jewellery try-on: Enhancing consumer engagement and personalization. *International Research Journal on Advanced Engineering Hub (IRJAEH)*, *2*(12), 2811–2815.

Gutiérrez, F., Verbert, K., & Htun, N. N. (2018). PHARA: An augmented reality grocery store assistant. *Proceedings of the 20th International Conference on Human-Computer Interaction with Mobile Devices and Services Adjunct*, 339–345.

Gutierrez-Antunez, A. R., & Leal-Enríquez, E. (2024). Augmented reality in retail: Metrics in the furniture department. *IAENG International Journal of Computer Science*, *51*(8), 1144–1154.

HappyFresh. (2015). *HappyFresh.* https://www.happyfresh.com/

Henriques, A. C., & Winkler, I. (2021). The advancement of virtual reality in automotive market research: Challenges and opportunities. *Applied Sciences*, *11*(24). https://doi.org/10.3390/app112411610

IKEA. (2017). *IKEA Place app launched to help people virtually place furniture at home*. https://www.ikea.com/global/en/newsroom/innovation/ikea-launches-ikea-place-a-new-app-that-allows-people-to-virtually-place-furniture-in-their-home-170912/

Jisp Bussiness Solutions (2022). Nisa Rolls out Jisp's Industry-first Scan & Save Technology Nationwide. https://www.jisp.com/post/nisa-rolls-out-jisps-industry-first-scan-save-technology-nationwide?utm

Jin, B., Kim, G., Moore, M., & Rothenberg, L. (2021). Consumer store experience through virtual reality: Its effect on emotional states and perceived store attractiveness. *Fashion and Textiles*, *8*(1), 19. https://doi.org/10.1186/s40691-021-00256-7

Ketoma, V. K., Schäfer, P., & Meixner, G. (2018). Development and evaluation of a virtual reality grocery shopping application using a multi-Kinect walking-in-place approach. *Intelligent Human Systems Integration: Proceedings of the 1st International Conference on Intelligent Human Systems Integration (IHSI 2018): Integrating People and Intelligent Systems, January 7–9, 2018, Dubai, United Arab Emirates*, 368–374.

Konarzewski, B., & Reiner, M. (2023). Augmented shopping: Virtual try-on applications in eyewear e-retail. *European Conference on Software Process Improvement*, 289–299.

Lee, D., Kim, H. K., & Park, J. (2024). Design and verification of universal virtual shopping store application. *Virtual Reality*, *28*(4), 168. https://doi.org/10.1007/s10055-024-01063-3

Liu, Y., Liu, Y., Xu, S., Cheng, K., Masuko, S., & Tanaka, J. (2020). Comparing VR- and AR-based try-on systems using personalized avatars. *Electronics*, *9*(11). https://doi.org/10.3390/electronics9111814

Manocha, R., Dharwal, M., Kumar, N., Bhowmik, D., & Yadav, V. K. (2024). Augmented reality virtual mirrors—The innovative drivers for consumer purchase decision-making. *Proceedings on Engineering*, *6*(4), 1561–1574.

Mekala, M., & Arun Ganessh, K. (2025). A study on consumer perceptions and experiences of VR-based virtual car showrooms. *International Journal of Research in Marketing Management and Sales*, *7*(1), 453–458.

Mystakidis, S. (2022). Metaverse. *Encyclopedia*, *2*(1), 486–497.

Nott, G. (2022). *M&S takes AR wayfinding app trial public*. https://www.thegrocer.co.uk/marks-and-spencer/mands-takes-ar-wayfinding-app-trial-public/663880.article

Ozturkcan, S. (2021). Service innovation: Using augmented reality in the IKEA place app. *Journal of Information Technology Teaching Cases*, *11*(1), 8–13.

Rayome, A. D. (2018). How Sephora is leveraging AR and AI to transform retail and help customers buy cosmetics. *TechRepublic*, *15*. https://steepnews24.wordpress.com/wp-content/uploads/2018/02/sephora_cover_story_final.pdf

Ricci, M., Evangelista, A., Di Roma, A., & Fiorentino, M. (2023). Immersive and desktop virtual reality in virtual fashion stores: A comparison between shopping experiences. *Virtual Reality*, *27*(3), 2281–2296.

Samy, T. M., Asham, B. I., Slim, S. O., & Abohany, A. A. (2025). Revolutionizing online shopping with FITMI: A realistic virtual try-on solution. *Neural Computing and Applications*, *37*(8), 6125–6144.

Shrestha, P., Kapali, S., Gautam, S., Pokharel, V., & Giri, S. (2025). 3D Reconstruction of Shoes for Augmented Reality. *ArXiv Preprint ArXiv:2501.18643*. https://arxiv.org/abs/2501.18643

Spinner, J. (2018). *Augmented reality brings pasta packaging to life*. https://www.packagingdigest.com/smart-packaging/augmented-reality-brings-pasta-packaging-to-life

Stecuła, K., Wolniak, R., & Aydın, B. (2024). Technology development in online grocery shopping—From shopping services to virtual reality, metaverse, and smart devices: A review. *Foods*, *13*(23), 3959.

Tan, Y.-C., Chandukala, S. R., & Reddy, S. K. (2022). Augmented reality in retail and its impact on sales. *Journal of Marketing*, *86*(1), 48–66.

Violante, M. G., Vezzetti, E., & Piazzolla, P. (2019). How to design a virtual reality experience that impacts the consumer engagement: The case of the virtual supermarket. *International Journal on Interactive Design and Manufacturing (IJIDeM)*, *13*, 243–262.

Wagaine, M. (2022). *9 augmented reality grocery shopping examples*. https://poplar.studio/blog/augmented-reality-grocery-shopping-examples/

Wolniak, R., Stecuła, K., & Aydın, B. (2024). Digital transformation of grocery in-store shopping-scanners, artificial intelligence, augmented reality and beyond: A review. *Foods*, *13*(18). https://doi.org/10.3390/foods13182948

Xu, B., Guo, S., Koh, E., Hoffswell, J., Rossi, R., & Du, F. (2022). ARShopping: In-store shopping decision support through augmented reality and immersive visualization. *2022 IEEE Visualization and Visual Analytics (VIS)*, 120–124.

Xue, L. (2022). *Designing effective augmented reality platforms to enhance the consumer shopping experiences*. Loughborough University.

Yahoo! (2017). *Check Out This Grocery App in Malaysia That Wants You to Shop with Virtual Reality*. https://sg.news.yahoo.com/check-grocery-app-malaysia-wants-163212797.html?guccounter=1&guce_referrer=aHR0cHM6Ly93d3cuZ29vZ2xlLmNvbS8&guce_referrer_sig=AQAAAJDifyds3hXk6kvfpJeEVhl_VtfFRo45vlTtv5f4DtY4FPr91Bmd_De2itWUqQQUJ5LypX5TZ81veM04XGciDbKfFUGCVHltLo75

Yew, J. Y. (2023). *ARGrocery: An augmented reality-assisted grocery shopping mobile application*. UTAR.

Zhang-Wu, X. (2024). *AR Technology in Jewellery Display*, Metropolia University of Applied Sciences, Helsinki. https://www.theseus.fi/handle/10024/820726

CHAPTER 4

Sport

4.1 VIRTUAL TRAININGS, EXERCISES, AND ACTIVITIES

Virtual reality (VR) technologies are opening up new possibilities in sports and physical activity. By putting on VR goggles, users are transported to a digital environment where they can train their chosen discipline, practice fitness, or engage in activities previously only available in real-world settings. This allows even those without access to a sports hall or field to participate in realistic sports training from the comfort of their own living room. VR enables simulation of sports such as golf, skiing, soccer, or rowing, providing users with a realistic athletic experience. Furthermore, VR is used to simulate sports training and fitness classes—giving them the opportunity to exercise, move, and compete without leaving home or accessing traditional sports facilities.

Below, a list of selected sport disciplines and activities that can be supported by VR and examples of VR applications is presented:

- Golf: Walkabout Mini Golf application
- Skiing: Alpine Ski VR application
- Soccer: VR Soccer Training application
- Fitness: Supernatural application
- Boxing: Virtual Knockout application
- Tennis: Sense Arena for Tennis
- Rowing/Kayaking: MarineVerse Kayaking HoloFit application

DOI: 10.1201/9781003643395-4

- Cycling – VirZOOM Bike application
- Goalkeeping – Goalkeeper VR Challenge application
- Climbing – The Climb 2 application
- Summer Sports (for example, swimming, beach volleyball, jet skiing, water basketball, cycling, frisbee, water polo, and more): All-In-One Summer Sports VR application

Each of the aforementioned sports apps allows users to freely choose their preferred sport and immediately begin their activity – without having to leave home, commute, or access specialized sports infrastructure. This allows VR users to regularly train anywhere, anytime, which is particularly important for people living in a fast-paced environment, students, multi-shift workers, or parents who often cannot afford to travel to fitness centres. Figure 4.1 shows an illustrative example of a person exercising – riding an exercise bike with VR goggles.

Immersive technologies make it possible to adjust workouts to each user's needs and goals. Beginners can start with simple exercises, while advanced users can choose harder and more complex ones. The intensity,

FIGURE 4.1 An illustrative example of a person exercising – riding an exercise bike with VR goggles.

pace, and type of activity can be easily changed. This makes training more flexible, personal, and motivating. More and more apps offer this option.

It is also worth emphasizing that the virtual environment allows for exercise even when weather conditions are unfavourable and sports infrastructure is unavailable – for example, in winter, during rain, or in areas without sports facilities. VR can eliminate many traditional barriers when it comes to the access to physical activity.

VR lets users train in places like mountains, beaches, or stadiums. This makes exercise more fun and helps people stay motivated. It is especially helpful for those who have little access to nature or good sports facilities. A virtual environment combines movement, play, and immersion. It can create positive emotions, improve mood, and build healthy exercise habits. There are studies in the literature showing that VR training is positively perceived by participants and encourages greater activity (Gani et al., 2023).

To fully benefit from virtual training, users can utilize a variety of devices that support movement and feedback:

- VR Headsets with Controllers: These are the basic components of a VR system. However, it's worth noting that sometimes controllers are unnecessary, as the user's hand movement is captured.
- Motion Platforms (omnidirectional treadmills): Allow for walking/running in place.
- Sports Equipment/Sports Simulators: For example, a stationary bike connected to VR, ergometers connected to VR, devices simulating winter or water sports.
- Haptic Gloves and Vests: Give a sense of virtual resistance and contact.
- Training Exoskeletons: Support movement or provide resistance to increase exercise intensity.
- Foot and Hand Trackers: Improve the quality of limb movement reproduction.
- Body-Mounted Sensors: For example, an arm sensor for throwing.
- Full Body Tracking Systems: Ensure precise movement reproduction in real-time VR.

- Biometric Sensors: For example, wristbands, belts, or smartwatches; they monitor heart rate and body performance during VR training.
- Spatial Tracking Cameras: Used to precisely determine the user's position in space.

With such devices, training is not limited to passive movement – the user actively exercises the body using realistic stimuli and feedback.

The use of VR in sports and virtual training is connected with many benefits. They include the following:

- Ability to Practice Anywhere: User can train even from home, without having to travel to a sports facility.
- Immediate and Flexible Training: User can start exercising whenever they want; they do not have to wait for classes.
- Increased Physical Activity: Instead of sitting in front of a screen, people move, which improves health and fitness.
- Full-Body Workout: Virtual exercises can engage the entire body.
- Motivation Through Gamification: VR applications and games often are connected with competition, progress, and fun.
- Training in Winter or in Bad Weather: VR allows for activity regardless of weather conditions.
- Personalization: The user chooses the discipline, intensity, and pace and can customize the training.
- Educational Aspects: Technique simulation, visual feedback, movement visualization, and improved motor skills.
- For Children and Adolescents: An attractive form of activity, an alternative to sitting at a computer.

However, virtual trainings also are connected with some limitations and risks. They include the following:

- Social Isolation: Using VR a user do not have to train with another person; this, in a long term, can lead to isolation.

- User Comfort: Prolong use of goggles can cause dizziness, headaches, eye strain, or nausea.
- Risk of Injury: Exercising alone (without the control of a trainer) can be resulted in technical mistakes or overloading the body.
- Limited Sensory Experience: VR technology does not fully replicate touch, resistance, or other real conditions; as a consequence, the acquired skills may not be of good quality.
- High Cost: Professional VR equipment can be expensive; not everyone can afford it.
- Space Requirements: Some VR applications and equipment require a lot of space; not every user has a possibility to have enough space for VR sports.

4.2 FROM WATCHING TO PARTICIPATING IN SPORT THROUGH AUGMENTED REALITY

AR technology is becoming an increasingly important element in the sports sector – not only in the context of training or competition, but also in the spectator experience. Thanks to digital overlays on sports broadcasts, apps utilizing smartphone or tablet cameras, and interactive AR devices, fans, players, and coaches gain new tools – both for analysing and participating in sporting events. This section will present two main areas of AR application in sports: first, AR in the context of watching sports and engaging fans; second, practical applications of AR in sports training and applications for athletes and amateurs.

During sports broadcasts or live events, AR can add an interactive visual layer to the screen. It may show extra data, animations, 3D objects, tactical lines, or player profiles in real time. This helps viewers understand the game better and see its strategy more clearly. A few years ago, such insights were available only to coaches and analysts. Thanks to AR, viewers can 'immerse themselves' in the match, as if they were closer to the pitch, court, or stadium. In popular sports such as soccer, basketball, American football, and tennis, AR overlays enable the presentation of virtual offside lines, precise ball trajectories, pressing zones, animated offensive plays, passing predictions, and data on player speed and positioning. Viewers receive not only an entertaining image but also a tool for understanding tactics and real-time 'reading' of what's happening on the pitch.

Examples of AR applications in sports viewing:

- Soccer: Virtual offside lines, ball trajectory, expected goal analysis, player movement maps, lineup presentations in the form of 3D holograms on the pitch before the game.
- American Football: The 'yellow line' first-down + tactical graphics showing the game plan and player positioning.
- Basketball: For example, in the National Basketball Association (NBA), overlays showing shot distances, player speeds, movement paths on the court, and analysis of defensive positions.
- Car Racing: For example, in Formula 1, real-time display of car telemetry data—speed, tire changes, time differences, driving trajectories.
- Tennis: 3D display of ball flight paths and impact locations.
- Baseball: Graphical cues regarding ball speed, impact angle, spin, and player statistics.
- Ice Hockey: Puck and its trajectory highlighting for improved visibility at high game speeds.
- Cycling: For example, the Tour de France, AR shows elevation gain, course gradient, competitor pace, and distance to the finish line.
- Boxing/MMA: Holograms displaying statistics on punch power, hits, and distance travelled in the Octagon.
- Golf: Ball trajectory, analysed shots, and a course map.
- E-sports: Virtual heroes and game elements appearing in the arena during the broadcast.

Recent study shows that AR greatly improves the experience of watching sports (Pickman, 2023). It increases viewer engagement and helps people better understand the flow of the game. Such an environment fosters a stronger emotional connection with the event and increased fan satisfaction. Today, AR is not just a new gadget – it is becoming an important tool that shapes how people watch sports. Researchers point out that more investment and deeper studies are needed. They can help sports organizations, broadcasters, and marketers meet audience expectations, improve viewing quality, and find new sources of income. Other studies

(Minaya-Isique et al., 2024) show that real-time AR tools – like tactical graphics or zone lines – are already being used in different countries and sports.

The benefits of this form of AR application are significant:

- Increasing the attractiveness of sports broadcasts (for viewers).
- The ability to personalize content (for example, selecting a specific player, statistics, tracking the action).
- New advertising and branding opportunities (sponsored overlays, virtual effects).
- Better understanding of tactics and game flow for less experienced fans.

At the same time, challenges must be recognized: the need for real-time data synchronization, ensuring the smoothness of overlays during dynamic action, as well as issues related to information overload and viewer data security.

The other area is the use of AR for sports training – both in professional sports and in amateur or fitness activities. AR can simulate specific sporting situations, analyse user's movement, and help to improve technique (Soltani & Morice, 2020). It also offers new, motivating forms of exercise.

Examples of AR applications in sports are:

- Golf: Golfshot (AR) – an application for golf; it analyses shots, identifies obstacles, and visualizes trajectory on a real course.
- Fitness: FYTER – AR Fitness Workouts – an application that tracks motion and visualizes obstacles to make training interactive.
- Team Sports: Sense Arena – VR/AR training for hockey and tennis players; improving reactions, perceptions, and sports decisions.
- Other Training Forms, Including Running: 'AR Runner' – an AR running game with checkpoints and competition in a real-world environment.
- Team Events: ARound – an AR solution for sporting events, stadiums, shared entertainment, and fan interaction.

FIGURE 4.2 An illustrative example of a person running with AR goggles – a person sees virtual checkpoints in the real world.

An interesting solution is training in AR goggles, which are used while running. The runner sees, for example, virtual checkpoints and this makes running more attractive because he or she has to pass the next checkpoint. Figure 4.2 shows an illustrative drawing of such a solution. Figure 4.3 shows a person playing golf with AR support.

FIGURE 4.3 An illustrative example of a person playing golf with AR support.

The main advantages of AR training are easy access and higher motivation. People can train anytime, no matter the weather. Gamification and quick visual feedback help users stay engaged. There is no need for special equipment or a gym, which lowers the barrier to exercise. This technology can also help increase physical activity in modern societies, where many people lead sedentary lives.

AR in sports transforms two key dimensions: the fan experience and the user's own athletic activity. Fans get interactive, personalized, and more exciting broadcasts. Athletes and amateurs gain new tools for better, more precise, and more enjoyable training. To use the full potential of AR, it is important to focus on comfort, user knowledge, and social balance. AR in sports is not just a toy – it is a way to push limits and bring more people into active participation.

4.3 THE POTENTIAL OF VIRTUAL AND AUGMENTED REALITY TO ACTIVATE SOCIETY

In an era of rapid digitization and technological change, societies face two challenges: on the one hand, an excess of passive forms of digital content consumption, and on the other, the growing problem of physical and social deactivation. Technologies such as VR and AR can face these issues. They can counteract these trends. They can be used to design new forms of activity which has following features: physical, social, and educational. This new type of activity can engage users and connect the real and digital world. It can support activity in smart cities and in line with Industry 4.0 concept.

Immersive technologies make physical activity more fun. They let users join virtual sports, fitness classes, AR challenges, or group VR sessions with friendly competition (Deng et al., 2024). This gives less active people a new and interesting way to move instead of sitting all day. Unlike traditional workouts that need a gym or travel, VR and AR let people exercise anywhere – at home, in a park, or in a smart city space. Such activation can have a preventative effect on health issues, improve well-being, and support social integration through shared virtual activities.

In the context of smart cities, VR and AR are tools for increasing resident activity, engaging them in urban projects, and improving quality of life. AR can be used in mobile applications for urban educational walks, collaborative virtual training in urban spaces, or interactive health campaigns. Some research (IEEE Innovation at Work, 2022) indicate that AR can improve urban mobility, increase safety, and promote public health. It is then worth designing digital-physical interactions that encourage users to leave their homes, engage in urban spaces, and collaborate with other residents.

It is also worth relating VR and AR technologies to the concept of 15-minute cities. This concept assumes that residents should have access to basic services (for example, education, health, recreation, culture) within a short walk or bike ride (Carvalho et al., 2025). In this model, augmented and virtual reality can be tools that further support local activity. AR can enrich urban exercise paths, parks, and playgrounds with virtual tasks, gamification elements, and interactive training instructions available practically 'just around the corner'. VR, on the other hand, can create an alternative where physical infrastructure is limited, allowing residents of smaller districts access to virtual gyms, fitness centres, group training sessions, and virtual sports competitions. In this way, immersive technologies can support the idea of locality, reduce the need for commuting, and simultaneously activate communities, promoting healthy lifestyles and strengthening local ties through digital-physical activity environments.

It's also worth emphasizing that virtual and augmented reality can become a modern form of sports activities for children, youth, and students. Using VR and AR in physical education can make sports activities more attractive. It refers especially to students who have no will or motivation to participate in traditional exercises. VR movement games, virtual fitness training, sport simulations, and interactive AR tasks can increase motivation to exercise, develop coordination, and improve physical fitness, while combining activity with elements of fun. Such solutions could particularly benefit students who are less able, shy, or have limited access to sports facilities, giving them the opportunity to participate in activities in a safe and engaging digital environment while also supporting the development of technological competencies.

One interesting area is the combination of e-sports, VR, and AR as a way to engage people who aren't typically interested in traditional sports. Imagine a neighbourhood VR/AR centre where users wearing goggles participate in virtual exercise competitions – for example, virtual athletics, dance, climbing, or team sports in AR. This allows players to 'train' and 'compete', while also developing movement, coordination, and health. This model can be effective when it comes to activating young people. It can refer to those with limited mobility but also – who prefer video games rather than traditional sports.

A vision for the future can be envisioned: a city implementing a program for physical activation in the city. Residents receive an AR app that displays activity points in the city (e.g., exercise benches, VR city bikes, AR dance stations). Simultaneously, VR sessions are launched at the local community centre – virtual fitness classes, VR training for seniors, and competitions for children. Such an initiative integrates the community, reduces the barrier to entry into activity, strengthens intergenerational bonds, and activates people in everyday urban spaces. In the scientific literature, there are studies on applications in smart cities, pointing to the potential of VR/AR in promoting social and educational activation (Xiang et al., 2023).

Of course, the potential of VR/AR in activating society does not mean there are not challenges. Above all, there's a risk that technology will replace real-world social interactions, which could lead to alienation. There are also real barriers to using these technologies. High costs, limited access to equipment, and competition with traditional sports make adoption harder. Health issues are also important – spending too much time in VR can cause tiredness or eye strain (Urbanek et al., 2025). In addition, movement in virtual spaces is not always the same as real physical activity. In smart cities, it is also important to ensure digital inclusion so that VR and AR programs do not increase social inequality. In short, success depends on connecting technology with real health and social goals, educating users, and creating solutions that inspire real action – not replace it.

BIBLIOGRAPHY

Carvalho, T., Farber, S., Manaugh, K., & El-Geneidy, A. (2025). Assessing the readiness for 15-minute cities: A literature review on performance metrics and implementation challenges worldwide. *Transport Reviews*, *45*(5), 801–827.

Deng, J., Zhang, Y., Chen, H., Wang, Z., & Tuo, S. (2024). VR-enhanced situational teaching with digital competitive gaming: Examining comprehensive physical fitness and psychological adaptations. *Proceedings of the 2024 International Conference on Sports Technology and Performance Analysis*, 494–501.

Gani, R. A., Setiawan, E., Achmad, I. Z., Aminudin, R., Purbangkara, T., & Hofmeister, M. (2023). Virtual reality-based tabata training: A professional method for changing levels physical fitness and psychological well-being on student-athletes. *Pedagogy of Physical Culture and Sports*, *27*(2), 91–101.

IEEE Innovation at Work. (2022). *How Will New Developments in Virtual and Augmented Reality Technology Impact Smart Cities?* https://innovationatwork.ieee.org/how-will-new-developments-in-virtual-and-augmented-reality-technology-impact-smart-cities/

Minaya-Isique, J., Gamboa-Cruzado, J., Núñez-Meza, Á, & López-Ramírez, B. C. (2024). Augmented reality and its influence on sporting events: A systematic literature review. *Computación y Sistemas*, *28*(4), 1961–1980.

Pickman, D. (2023). The use of virtual reality and augmented reality in enhancing the sports viewing experience. *International Journal of Arts, Recreation and Sports*, *1*(2), 39–49.

Soltani, P., & Morice, A. H. P. (2020). Augmented reality tools for sports education and training. *Computers & Education*, *155*, 103923.

Urbanek, A., Stecuła, K., Kaźmierczak, K., Łagosz, S., Kwoczak, W., & Dyczko, A. (2025). Towards sustainable health and safety in mining: Evaluating the psychophysical impact of VR-based training. *Sustainability*, *17*(13), 6205.

Xiang, M., Wang, R.-H., Zhong, Z., Hong, C. Y., Yi, Y. W., Xiang, S. W., & Roy, D. (2023). Empowering society 5.0 through AR-VR applications: Studying the potential in Aizuwakamatsu smart city. *AIP Conference Proceedings*, *2909*(1), 70007.

CHAPTER 5

Tourism

5.1 VIRTUAL TOURISM: EXPLORING THE WORLD WITHOUT LEAVING HOME

The development of immersive technologies such as virtual reality (VR) is opening up entirely new possibilities in the field of tourism. Virtual tourism allows users to transport themselves to distant places, cities, museums, and national parks without the need for physical travel. Using VR goggles, controllers, and specific VR application, visitors can explore realistic recreations of spaces, stroll along historic streets, visit the interiors of historical buildings, and even experience adventures that would be difficult or impossible in the real world. Virtual tourism is not merely an alternative to traditional forms of travel, but is developing as a parallel field of cultural, educational, and recreational experiences. It facilitates access to world heritage for those who, for various reasons – health, financial, or logistical – cannot travel long distances. Some studies discuss VR's capacity to use tourism experience for, for example, seniors and disabled people who face barriers to physical travel (Iftikhar et al., 2023; Sirih et al., 2025). Furthermore, VR enables tourism and cultural institutions to create innovative forms of promotion (Ozdemir, 2021), attracting new audiences and increasing interest in destinations (An et al., 2021; Ying et al., 2022).

There are many applications available on the market that enable virtual tours of monuments, cities, museums, and natural sites. Depending on the user's needs, these can be completely free or paid versions offering more extensive experiences, higher-quality graphics, professional guidance, and educational features. Such programs can be downloaded from international software distribution platforms such as SteamVR,

DOI: 10.1201/9781003643395-5

Meta Store, Viveport, or Google Play/Apple App Store, as well as from the websites of individual cultural institutions, museums, and tourist centres, which are increasingly offering their own immersive applications. One of the examples is VR application called 'Mona Lisa: Beyond the Glass'. It is the Louvre's first VR project which uses the latest scientific research on Leonardo da Vinci, his creative processes, and his painting techniques, as it can be read on the website (Louvre.fr, 2021). The application begins in the Louvre, bringing the user face-to-face with the famous Mona Lisa. Then, they are transported through time and space. A virtual loggia, inspired by Leonardo da Vinci's sketches, opens behind the figure to reveal a panoramic landscape seen in the painting's background. In this space, they can take a closer look at the painting's techniques: subtle sfumato, layering of light and shadow, as well as details that are usually invisible – such as hidden sketches, X-ray analysis, and Leonardo's techniques. The story of the portrait itself is explored – who the model (Lisa Gherardini) may have been and how the work was created.

What is more, VR users can explore various places around the world, including tourist destinations, for example, they can see Eiffel Tower (Viality Tour, 2025), Sistine Chapel ('IL DIVINO: Michelangelo's Sistine Ceiling in VR' application) (Steampowered, 2019), Notre Dame ('Notre-Dame de Paris: Journey Back in Time' application) (Steampowered, 2020), Great Pyramid ('Great Pyramid VR' application) (Steampowered, 2017), or even national parks (Meta, 2025). They can reach inaccessible spots such as the underwater world – for example, due to the Ecosphere application (Meta, 2020) – or the surface of planets – for example, Mars (MarsVR, 2023). Figure 5.1 shows an illustrative example of a person virtually visiting the Eiffel Tower.

In addition to VR apps for visiting a specific museum or monument, there are applications that allow a user to visit any place in the world. VR applications allow users to virtually travel to almost any location in the world without having to travel. Users can walk along selected streets, enter historic buildings, and view monuments and structures from different perspectives. Users can explore their surroundings, zoom in on details, switch between locations, and even interact with their surroundings. Examples of such applications are (Stecuła & Naramski, 2025) Wander (Meta, 2019); Google Earth VR (Steampowered, 2016); National Geographic Explore VR (*National Geographic Explore VR*, 2019); BRINK Traveler (Steampowered, 2021); Sites in VR (Ercan Gigi, 2016); Wooorld (Meta, 2022); and more.

FIGURE 5.1 An illustrative example of a person virtually visiting the Eiffel Tower using virtual reality headset.

Many applications now include stories, historical scenes, and virtual guides that make exploring more fun and educational (Kontogiorgakis et al., 2024). Users can explore sites on their own or follow ready-made tours with voice narration, short texts, and videos. Shared tour platforms are also becoming popular.

They let people visit the same place together with friends, family, or a live guide. This makes travel a social experience, even though it takes place in a digital space. An example is the Hoppin' VR application (Hoppin' World, 2021). This is a social platform for multiple users. It offers virtual discovery of new places – but not just solo. It's possible to view and explore selected locations simultaneously with other platform users. For example, a sales representative can give a virtual tour of a potential client. A teacher can lead a virtual trip. A family can also explore a location together using VR goggles. Users share the same space and interact. The application offers private shared spaces. This solution allows users to discuss what they see in real time, also building personal relationships.

VR in tourism offers a number of benefits that make it a modern, ecological, and inclusive form of travel. Above all, VR significantly reduces the negative impact of tourism on the environment. Virtual travel reduces the need for physical travel, reducing CO_2 emissions, air pollution, and the burden on ecosystems exposed to excessive tourism. This technology can help in the protection of environmentally and culturally sensitive sites. What is more, it can support the concept of sustainable and responsible tourism (Miłosz et al., 2025; Su et al., 2024).

VR is also an educational tool (Hussein & Nätterdal, 2015; Zouboula et al., 2008). It lets users to visit monuments, museums, and other places of the world without travelling. It offers engaging experience. By enabling the reconstruction of historical events, realistic representations of nature, and guided tours by virtual guides, VR serves as an educational platform that combines learning with emotion and interactivity.

Another significant advantage is the accessibility of this technology to people who, for health, financial, or logistical reasons, cannot travel (Stecuła & Naramski, 2025). Virtual tours allow seniors, people with disabilities, and residents of remote areas to explore the world without barriers, which promotes the democratization of tourism and equal opportunities in accessing culture.

VR technology also creates new business opportunities. It helps promote less-known regions, build tourism content, and develop digital guide services. It also lets tourists explore their destination before travelling. This increases satisfaction and makes planning easier (Beck et al., 2019).

VR can also support mental well-being. It helps people relax and escape daily stress (Fagernäs et al., 2021; Verstegen et al., 2025). It offers realistic natural scenes, sea beaches, and exotic landscapes that support regeneration and stress reduction (Gentile et al., 2023).

In summary, virtual tourism connects sustainability, education, and accessibility. It makes travel more conscious and eco-friendly, while keeping the joy of discovery. Table 5.1 presents main benefits of visiting places and playing tourism applications in VR.

VR has many benefits for tourism, but it also has limits. VR cannot give the full feeling of real travel. It shows realistic images, but you cannot smell the sea, feel the forest air, or touch surfaces. Without smell, taste, and touch, the experience feels less real and less emotional. New tools, like haptic gloves, may help in the future. They could let users feel different textures in virtual worlds (Perret & Vander Poorten, 2018; Yang et al., 2021).

VR also cannot give true cultural contact. Meeting locals, eating local food, hearing the language, and joining customs are key parts of travel. Virtual tourism, even if immersive, cannot replace being physically present. So, VR can support tourism, but it cannot replace real travel when deep cultural experience is the goal.

Additionally, a sudden big shift to virtual tourism would have negative consequences. It would be economic threat to many regions dependent on the tourism (Vinnakota et al., 2023). This industry generates millions of jobs in the hotel, restaurant, transportation, and tour guide sectors.

TABLE 5.1 Main Benefits of VR Tourism Activity

Category	Key Benefits	Description
Environmental Benefits	Reduced CO_2 emissions	Less physical travel means fewer greenhouse gas emissions from air and road transport.
	Lower environmental strain	It helps protect nature and historical sites from overtourism, waste, and erosion.
	Conservation support	It enables visitors to explore delicate ecosystems (reefs, polar regions, deserts) without damaging them.
	Sustainable tourism education	Promotes eco-awareness and responsible travel behaviour.
Educational and Cultural Benefits	Access to global heritage	Users can explore museums, monuments, and landmarks from anywhere.
	Immersive learning	VR allows for engaging and interactive lessons in history, science, and geography.
	Cultural preservation	Digital archives help preserve cultural heritage for future generations.
	Innovative storytelling	Virtual guides, reconstructions, and multimedia enhance knowledge retention.
Social and Accessibility Benefits	Inclusivity for all	People with health, financial, or mobility limitations can 'travel'.
	Safe exploration	Suitable for seniors and vulnerable users.
	Democratization of tourism	Everyone can visit global destinations regardless of income or location.
	Social VR travel	Travelling together with family, friends, or an online guide is possible.
Technological and Economic Benefits	Promotion of lesser-known destinations	It helps local communities gain exposure without developing tourism infrastructure.
	New business opportunities	Growth of VR industry: 3D design, app development, online guides.
	Pre-travel simulation	Travellers can preview destinations, hotels, planned trips.
	Remote revenue streams	Museums and attractions can sell virtual tours.

(Continued)

TABLE 5.1 Main Benefits of VR Tourism Activity *(Continued)*

Category	Key Benefits	Description
Psychological and Recreational Benefits	Stress reduction	Virtual environments of nature and travel support relaxation and mental health.
	Sense of escape	They provide an emotional escape and a sense of 'escape' to another place.
	Realistic experiences	Users feel present in exotic landscapes and unique locations.
	Gamified exploration	Motivation through achievements, quests, or guided missions.
Innovation and Future Potential	Hybrid tourism model	Combines VR with tourism in the real world.
	Educational tourism evolution	Supports global learning programs, universities, online schools.
	Accessible cultural diplomacy	Nations can showcase culture worldwide through immersive media.
	Technological advancement	They drive innovation in haptics, 360° video, and virtual avatars.

Source: Author's own work based on Stecuła and Naramski (2025).

Replacing physical travel with virtual travel could have negative economic consequences, especially in countries where tourism is a key economic pillar. Although the development of VR creates new jobs in the technology industry, this transformation requires time, investment, and costly employee training.

It is also worth noting the potential impact of VR on social relationships and individual functioning. Virtual experiences are often experienced in isolation, which can foster feelings of loneliness, alienation, and a gradual reduction in interpersonal contact. This phenomenon can exacerbate the problem of technology addiction. The issue of VR addiction is a matter of research already (Rajan et al., 2018). To much use of VR can lead to the will to escape from reality. What is more, it can cause emotional disturbances, and health problems such as insomnia, headaches, and eyestrain (Ülker et al., 2024; Urbanek et al., 2024). There are also even dissociative symptoms mentioned in the literature (Taveira et al., 2022). Furthermore, excessive digital presence can lead to a gradual blurring of the lines between the real and virtual worlds, which over time can weaken empathy and the ability to build lasting social relationships. There is a risk that some users will begin to prefer the simulated world over real experiences, potentially leading to a closed society.

Financial and technological barriers are another limitation (Atkins & De Wilde, 2017). To have a good quality VR session, users must buy advanced VR equipment and have a fast internet connection. These things can be a challenge for some people. Furthermore, VR requires constant technological development. Technical errors or poor graphic quality can reduce the level of immersion a lot.

In summary, VR is a modern, valuable, and ecological tool supporting tourism. But its widespread implementation brings many challenges. Therefore, it is crucial to develop VR while taking into account social, economic, and health aspects, striving for a balance between technology and real-world travel experiences. Table 5.2 presents main drawbacks of visiting places and playing tourism applications in VR.

TABLE 5.2 Main Drawbacks of VR Tourism Activity

Category	Key Drawback	Description
Lack of Full sensory Immersion	VR cannot replicate smells, tastes, natural tactile sensations, or climatic conditions; this limits authenticity.	No scent of the sea, wind, temperature, or ability to touch objects.
Limited Cultural Authenticity	Lack of real interaction with local communities, traditions, and atmosphere; VR cannot replace cultural immersion.	No contact with locals, cuisine, language, customs.
Risk to Tourism-Dependent Economies	Replacing physical tourism may negatively affect regions reliant on travel income.	Reduced revenue for hotels, restaurants, airlines, guides.
New Workforce Requirements	Industry transformation requires training, investment, and time; some jobs may disappear.	Guides needing VR skills; time needed to adapt industry.
Risk of Social Isolation	Excessive VR use may weaken social relationships, causing loneliness and alienation.	Preference for virtual worlds over real interaction.
Potential Technology Addiction	VR, like social media and games, may cause addictive behaviours and psychological effects.	Escaping into VR instead of dealing with real life; digital escapism.
Health-Related Risks	VR may cause physical and psychological discomfort.	Motion sickness, eye strain, headaches, sleep disruption.

(Continued)

TABLE 5.2 Main Drawbacks of VR Tourism Activity *(Continued)*

Category	Key Drawback	Description
Economic Barrier for Users	High-quality VR systems are expensive and require modern technology and fast internet.	Cost of VR headsets, subscription apps, fast internet needs.
Reduced Physical Activity	Virtual tourism replaces real travel and limits physical benefits.	No walking, hiking, contact with nature.
Technical and Quality Limitations	Realism depends on hardware, software, and connectivity; bugs can reduce immersion.	Lagging apps, lower resolution, software glitches.
Ethical and Societal Concerns	Deep immersion in VR may replace real engagement and reduce empathy for real-world issues.	Preferring virtual experiences over real ones; detachment from real communities.

Source: Author's own work based on Stecuła and Naramski (2025).

5.2 AUGMENTED REALITY ENHANCING TRAVEL EXPERIENCES

AR technology is more and more popular in the field of tourism and travel. It is not as a backup environment like VR, but as a reinforcement of real-world sightseeing. Thanks to AR, users can receive additional information on their smartphone screens, laptops, or through AR glasses during city walks, museum visits, or excursions to natural sites: visualizations, animations, layers of history, tips, and interactive elements that overlay the surrounding reality.

When visiting monuments, cities, or exhibitions, users can point their device's camera at a building, work of art, or exhibit, and the AR app will display information in the form of text, images, 3D models, or animations (Spadoni et al., 2022; Vargün, 2023). This makes the tour more interactive and educational. Visitors can see how a given structure looked like in the past. They can see, for example, direction of windows or portals, the functions of buildings, or the social context. AR becomes a kind of guide. It accompanies the user and guides the tour. AR shares interesting facts and helps to understand the history, art, and geology (Tsepapadakis & Gavalas, 2023). In practice, AR can fulfill many specific functions. Table 5.3 shows examples of AR usage.

As mentioned in the table, one of the examples of using AR is application which works as follows: a user places a tablet or smartphone over a given monument, and they can see digitally reconstructed version of the monument. Figure 5.2 shows an illustrative example of using AR to see the digitally reconstructed version of the colosseum.

TABLE 5.3 Examples of AR Usage in Tourism

Field of AR Usage	Function Description	Example + Source
Navigation and Orientation	AR applications provide real-time directions, display arrows to attractions, overlay the route on the sidewalk, and mark landmarks.	Sightseeing Support System (Japan) – an AR system that shows tourists directions in urban space using arrows and pictograms, supporting orientation in the field (Sasaki & Yamamoto, 2019).
Overlays on Buildings and Historical Sites	AR allows a user to look at buildings and historical places with layers of information superimposed: former appearance, timeline of changes, reconstructions of destroyed elements, animations.	AR-Locations – a solution presenting reconstructions of historical buildings and showing what they looked like in the past by overlaying digital layers on the real view (AR Locations by 3DQR, 2023).
Interactive Exhibits in Museums	Visitors point their smartphone at the exhibit and receive additional content: animations of the work being created, curator comments, 3D models, and interactive descriptions.	MuseumNext – the paper on the webside of MuseumNext describes AR projects in museums, including image animations, 3D reconstructions of objects, and interactive visitor guides (MuseumNext, 2025).
Maps and Physical Plans with AR	After scanning a physical map or model with your phone, 3D building models, animations, and suggested sightseeing routes are displayed.	EuphoriaXR – AR models and maps – the system enables scanning of physical maps that come to life through 3D models and urban animations supporting tourist orientation (EuphoriaXR, 2025).
Gamified and Collaborative Exploration (Quests, Field Games)	AR enables collaborative exploration, field games, problem-solving, group collaboration, and interaction with markers and QR codes.	ToARist application – an AR travel application designed using user-centred design, enabling interactive, social tours with gaming elements (Williams et al., 2018).

Source: Author's own work.

Literature and practice provide specific examples of AR being used in tourism and sightseeing. Examples include:

- Latvian National Museum of Art (Latvia): A mobile application with an AR module was created that allows visitors to point their smartphone at a work of art and receive additional visual information, thus addressing the lack of guides (Overly, 2022).

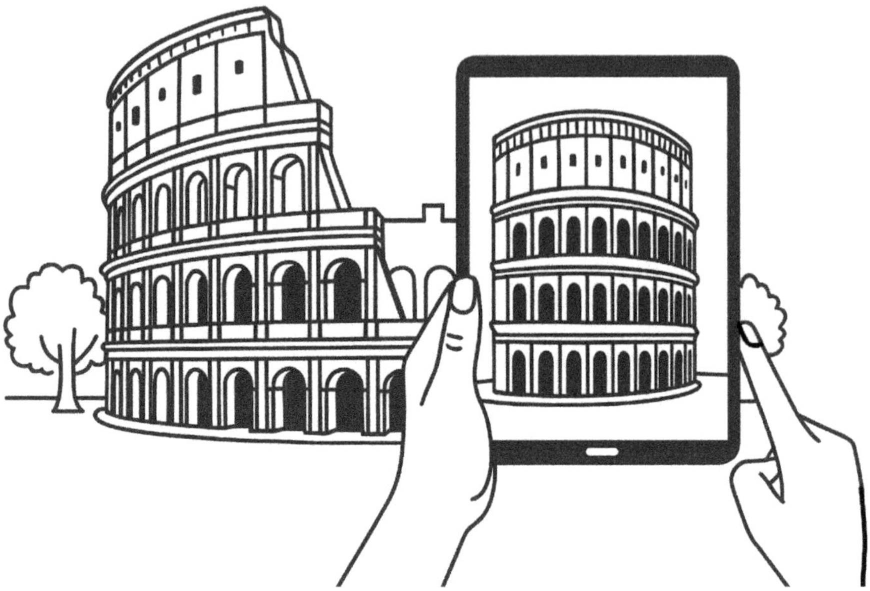

FIGURE 5.2 An illustrative example of using AR to see the digitally reconstructed version of the colosseum.

- Museum of Applied Art (MAK) in Vienna: In collaboration with an AR company, Gustav Klimt's The Tree of Life was presented in an AR version, enabling animation and interaction with it (Cuseum, 2025).
- AR technology was used to create an interactive museum app that superimposed three-dimensional animated elements onto a fragment of the Chinese painting 'Along the River During the Qingming Festival'. Visitors could scan the painting with a tablet, view the three-dimensional scene, zoom in and rotate objects, and obtain additional information by touching selected elements. This made the experience more engaging (Gong et al., 2022).
- An AR application has been developed in China's Mogao Caves: It allows visitors to see reconstructed walls and frescoes and explore the history (Li & Kieanwatana, 2024).
- Industrial Museum and Cultural and Educational Center in Thessaloniki (Greece): A 'True AR' case study at the Industrial Museum in Thessaloniki. There, digital objects were placed on real exhibits to see if an 'almost indistinguishable' experience enhanced the visitor experience (Virtual Multimodal Museum, 2020).

AR offers numerous benefits for the tourism sector, museums, and cultural heritage sites. AR enhances traditional sightseeing. It combines the real world and digital information. This makes travel more engaging, interactive, and educational. The key advantages of using AR in tourism include:

- Increasing user engagement through interactive content, animations, and 3D reconstructions.
- Better understanding of history, art, and architecture through visualizations and contextual narratives.
- The ability to 'see the past' – reconstructing monuments and recreating the appearance of non-existent elements.
- Personalizing the visitor's path and tailoring content to the user's interests.
- Access to multilingual educational materials and content for people with disabilities (e.g., voice guides, subtitles).
- Eliminating the need to use traditional guides or paper maps.
- Gamification of the tour experience (quests, rewards, puzzles).
- Easy updating and expanding content without interfering with actual exhibits.
- Creating more attractive and modern tourism offers that strengthen the competitiveness of regions.
- Supporting educational and family tourism – increased motivation for children and young people to learn about history.
- Potential for integration with AI and recommendation systems to dynamically tailor content to the user.

Despite benefits, AR in tourism also has drawbacks and challenges. The technology can distract from the surroundings. It requires appropriate equipment and internet access. Also, content creation is expensive and time-consuming. Furthermore, AR applications more and more often use AI. This, in turn, can generate inaccurate or false information. As a consequence, it is a big risk for the accuracy of educational content. The main limitations of AR in tourism are:

- Distraction and reduced focus on the real environment.
- Hardware dependence (smartphone/tablet/AR glasses).
- Internet dependence.
- Technological barrier for less advanced users.
- High cost of creating and maintaining AR content (3D models, animations, localization systems).
- Possible localization errors and inaccurate AR layering.
- Risk of rapid battery drain on mobile devices.
- AI modules may generate erroneous or fabricated content (so-called AI hallucinations) – a threat to historical education and the reliability of information.
- Potential privacy issues (user location, image analysis).
- A certain 'addiction' to technology instead of natural, spontaneous sightseeing.
- Prolonged use of AR goggles can cause discomfort, headaches, eye pain, dizziness, and other ailments.

5.3 OPPORTUNITIES AND POSSIBILITIES OF VIRTUAL AND AUGMENTED REALITY IN TOURISM

VR and AR are opening up new possibilities in the tourism sector, transforming travel planning, sightseeing, and cultural education. Both technologies, despite the fact that they are operating in different environments (VR creates a completely virtual world, while AR overlays digital elements onto reality), are important tools supporting the development of modern, sustainable, and digitally accessible tourism. Opportunities and possibilities of VR and AR in the mentioned field can be as follows:

- New Travel and Learning Experiences: VR lets people see realistic versions of places without leaving home. User can visit monuments, national parks, or museums from anywhere in the world. AR adds extra details to real trips, like animations, reconstructions, or interactive elements. This helps travellers understand history, culture, and architecture better, making trips more fun and interesting.

- Supporting Sustainable Tourism: VR can lower the number of visitors in fragile natural areas and monuments, helping protect them. Virtual tours are a safe option for places like coral reefs, old ruins, or nature reserves. These tools also make travel easier for people with health, money, or location limits. In this way, VR and AR support more inclusive tourism.
- Destination Promotion and Development of Local Economies: VR and AR are increasingly being used as promotional tools by cities, regions, and cultural institutions. They allow for the presentation of attractions before a trip. This can make tourists interested in future visits. Also, the development of these technologies generates new jobs. For example, there is a chance for work for 3D designers, programmers, graphic designers, digital guides, and educational content creators.
- Enhancing Museum and Cultural Offers: Cultural institutions are increasingly implementing VR and AR to recreate past realities, present reconstructions of destroyed structures, and provide access to exhibits not normally accessible. This allows museums and galleries to offer audiences modern, immersive experiences, increasing engagement, and motivation to learn.

In summary, VR and AR represent the future of the tourism industry, offering innovative forms of sightseeing, education, and destination promotion. On one side, VR and AR let people visit faraway or hard-to-reach places. This helps protect heritage and supports sustainable tourism. On the other side, they make real trips more fun and interactive. Their dynamic development creates vast opportunities for the tourism sector, both in terms of user experience and new business models.

BIBLIOGRAPHY

An, S., Choi, Y., & Lee, C.-K. (2021). Virtual travel experience and destination marketing: Effects of sense and information quality on flow and visit intention. *Journal of Destination Marketing & Management, 19*, 100492.

AR Locations by 3DQR. (2023). *Reviving the Echoes of the Past: Augmented Reality in Historical Site Restoration.* https://ar-locations.com/en/reviving-the-echoes-of-the-past-augmented-reality-in-historical-site-restoration/

Atkins, J., & De Wilde, P. (2017). Bridging the gap between reality and virtuality: An analysis of VR technology adoption within architectural practices. *Master's Thesis, University of Plymouth.*

Beck, J., Rainoldi, M., & Egger, R. (2019). Virtual reality in tourism: A state-of-the-art review. *Tourism Review, 74*(3), 586–612.

Cuseum. (2025). *Cuseum Partners with Pioneers & MAK to Celebrate Klimt with Augmented Reality.* https://cuseum.com/klimt-ar

Ercan Gigi. (2016). *Sites in VR.* http://www.sitesinvr.com/

EuphoriaXR. (2025). *Augmented Reality in Travel and Tourism: The Future of Interactive Travel Experiences.* https://euphoriaxr.com/augmented-reality-in-travel-tourism/

Fagernäs, S., Hamilton, W., Espinoza, N., Miloff, A., Carlbring, P., & Lindner, P. (2021). What do users think about virtual reality relaxation applications? A mixed methods study of online user reviews using natural language processing. *Internet Interventions, 24,* 100370.

Gentile, A., Ficarra, S., Thomas, E., Bianco, A., & Nordstrom, A. (2023). Nature through virtual reality as a stress-reduction tool: A systematic review. *International Journal of Stress Management, 30*(4), 341.

Gong, Z., Wang, R., & Xia, G. (2022). Augmented reality (AR) as a tool for engaging museum experience: A case study on Chinese art pieces. *Digital, 2*(1), 33–45. https://doi.org/10.3390/digital2010002

Hoppin' World. (2021). *Hoppin' VR. Step Inside. Together.* https://hoppin.world/platform/hoppin-vr

Hussein, M., & Nätterdal, C. (2015). *The benefits of virtual reality in education-A comparision Study, Department of Computer Science and Engineering Chalmers University Of Technology University Of Gothenburg Göteborg, Sweden.* https://scispace.com/pdf/the-benefits-of-virtual-reality-in-education-a-comparision-3omwkbqdih.pdf

Iftikhar, R., Khan, M. S., & Pasanchay, K. (2023). Virtual reality tourism and technology acceptance: A disability perspective. *Leisure Studies, 42*(6), 849–865.

Kontogiorgakis, E., Zidianakis, E., Kontaki, E., Partarakis, N., Manoli, C., Ntoa, S., & Stephanidis, C. (2024). Gamified VR storytelling for cultural tourism using 3D reconstructions, virtual humans, and 360° videos. *Technologies, 12*(6). https://doi.org/10.3390/technologies12060073

Li, J., & Kieanwatana, K. (2024). The impact of augmented reality tourism on tourist experience and behavior—A case study of the Mogao caves cultural heritage. *International Journal of Sociologies and Anthropologies Science Reviews, 4*(5), 103–122.

Louvre.fr. (2021). *The Mona Lisa in virtual reality in your own home.* https://www.louvre.fr/en/explore/life-at-the-museum/the-mona-lisa-in-virtual-reality-in-your-own-home

MarsVR. (2023). *Static Environment MDRS 360.* https://marsvr.com/

Meta. (2019). *Wander.* https://www.meta.com/pl-pl/experiences/wander/2078376005587859/?srsltid=AfmBOorZ8IxYaT0zjLcOJHzvXHmz-CsXCK2cY7KLKCFWAwq-3J3CRJGK

Meta. (2020). *Ecosphere*. https://www.meta.com/pl-pl/experiences/ecosphere/2926036530794417/

Meta. (2022). *Wooorld*. https://www.meta.com/pl-pl/experiences/wooorld/4360608937312293/?srsltid=AfmBOoqHGr5poo4UJy0Dq7EBZUnWGMzxOKavDxwFfGpjSRuJKi3tuUqB

Meta. (2025). *National Park Experience*. https://www.meta.com/pl-pl/experiences/national-park-experience/1524331530931445/?srsltid=AfmBOop8U5F1_omjow2_N4C6cmxwo1-k146iAgULT9uhuHH7gxPU6pXX

Miłosz, M., Żyła, K., Skulimowski, S. P., Dakowicz, A. L., Szymczyk, T., & Badurowicz, M. (2025). Virtual reality in supporting the creation of sustainable tourism: A case study of Gen Z technology acceptance. *Sustainability, 17*(16). https://doi.org/10.3390/su17167173

MuseumNext. (2025). *How Museums Are Using Augmented Reality*. https://www.museumnext.com/article/how-museums-are-using-augmented-reality/

National Geographic Explore VR. (2019). https://www.meta.com/pl-pl/experiences/national-geographic-explore-vr/2046607608728563/?srsltid=AfmBOoof2uolcUeGQt6lCJTQSD0qISkplAPIVZGeDQwoDBDxXoVY-7_j

Overly. (2022). *Museum offers augmented reality exploration for engaging storytelling*. https://overlyapp.com/case-study/museum-offers-augmented-reality-exploration-for-engaging-storytelling/

Ozdemir, M. A. (2021). Virtual reality (VR) and augmented reality (AR) technologies for accessibility and marketing in the tourism industry. In: *ICT tools and applications for accessible tourism* (pp. 277–301). IGI Global.

Perret, J., & Vander Poorten, E. (2018). Touching virtual reality: A review of haptic gloves. *ACTUATOR 2018; 16th International Conference on New Actuators*, 1–5.

Rajan, A. V., Nassiri, N., Akre, V., Ravikumar, R., Nabeel, A., Buti, M., & Salah, F. (2018). Virtual reality gaming addiction. *2018 Fifth HCT Information Technology Trends (ITT)*, 358–363.

Sasaki, R., & Yamamoto, K. (2019). A sightseeing support system using augmented reality and pictograms within urban tourist areas in Japan. *ISPRS International Journal of Geo-Information*, 8(9). https://doi.org/10.3390/ijgi8090381

Sirih, M. S., Duli, A., & Achmad, M. (2025). Exploring the potential of virtual reality for accessible tourism: The case of elderly and disabled in Makassar City, Indonesia. *Tourism Recreation Research*, 1–17. https://www.tandfonline.com/doi/full/10.1080/02508281.2025.2509177

Spadoni, E., Porro, S., Bordegoni, M., Arosio, I., Barbalini, L., & Carulli, M. (2022). Augmented reality to engage visitors of science museums through interactive experiences. *Heritage, 5*(3), 1370–1394.

Steampowered. (2016). *Google Earth VR*. https://store.steampowered.com/app/348250/Google_Earth_VR/?l=polish

Steampowered. (2017). *Great Pyramid*. https://store.steampowered.com/app/625190/Great_Pyramid_VR/

Steampowered. (2019). *IL DIVINO: Michelangelo's Sistine Ceiling in VR*. https://store.steampowered.com/app/1165850/IL_DIVINO_Michelangelos_Sistine_Ceiling_in_VR/

Steampowered. (2020). *Notre-Dame de Paris: Journey Back in Time.* https://store.steampowered.com/app/1341280/NotreDame_de_Paris_Journey_Back_in_Time/

Steampowered. (2021). *BRINK Traveler.* https://store.steampowered.com/app/1462520/BRINK_Traveler/?l=polish

Stecuła, K., & Naramski, M. (2025). Virtual reality as a green tourism alternative: Social acceptance and perception. *Sustainability, 17*(17), 7722.

Su, Z., Lei, B., Lu, D., Lai, S., & Zhang, X. (2024). Impact of ecological presence in virtual reality tourism on enhancing tourists' environmentally responsible behavior. *Scientific Reports, 14*(1), 5939.

Taveira, M. C., de Sá, J., & da Rosa, M. G. (2022). Virtual reality-induced dissociative symptoms: A retrospective study. *Games for Health Journal, 11*(4), 262–267.

Tsepapadakis, M., & Gavalas, D. (2023). Are you talking to me? An audio augmented reality conversational guide for cultural heritage. *Pervasive and Mobile Computing, 92*, 101797.

Ülker, S. V., Sümer, B. N., Sönmez Kence, E., & Hızlı Sayar, F. G. (2024). Psychophysiological investigation of the effects of virtual reality, the new dimension of retail shopping, on Generation Z. *International Journal of Human–Computer Interaction, 41*(7), 1–14.

Urbanek, A., Stecuła, K., Kaźmierczak, K., Łagosz, S., Welenc, H., & Dyczko, A. (2024). Virtual reality-based training for miners: An assessment from participants' perspective. *Scientific Papers of Silesian University of Technology. Organization & Management/Zeszyty Naukowe Politechniki Slaskiej. Seria Organizacji i Zarzadzanie, 197*(197), 615–628.

Vargün, Ö. (2023). Visual analysis of 3D characters and animations used in the presentation of cultural heritage at museums. *Journal of Arts, 6*(2), 141–155.

Verstegen, A., Van Daele, T., Bonroy, B., Debard, G., Sels, R., van Loo, M., & Bernaerts, S. (2025). Designing a smartphone-based virtual reality app for relaxation: Qualitative crossover study. *JMIR Formative Research, 9*(1), e62663.

Viality Tour. (2025). *Viality Tour.* https://vialitytour.com/en/

Vinnakota, S., Mohan, M. D., Boda, M. J., Askarzai, W., Devkota, M. P., Shetty, M. S., Wangmo, M. T., & Choden, M. T. (2023). Venturing into virtuality: Exploring the evolution, technological underpinnings, and forward pathways of virtual tourism. *Educational Research (IJMCER), 5*(6), 8–49.

Virtual Multimodal Museum. (2020). *Case Study for True Augmented Reality (AR) Application in Industrial Museum and Cultural and Educational Center in Thessaloniki.* https://www.vi-mm.eu/project/case-study-for-true-augmented-reality-ar-application-in-industrial-museum-and-cultural-and-educational-center-in-thessaloniki/

Williams, M., Yao, K. K. K., & Nurse, J. R. C. (2018). ToARist: An augmented reality tourism app created through user-centred design. *ArXiv Preprint ArXiv:1807.05759.* https://arxiv.org/abs/1807.05759

Yang, T., Kim, J. R., Jin, H., Gil, H., Koo, J., & Kim, H. J. (2021). Recent advances and opportunities of active materials for haptic technologies in virtual and augmented reality. *Advanced Functional Materials*, *31*(39), 2008831.

Ying, T., Tang, J., Ye, S., Tan, X., & Wei, W. (2022). Virtual reality in destination marketing: Telepresence, social presence, and tourists' visit intentions. *Journal of Travel Research*, *61*(8), 1738–1756.

Zouboula, N., Fokides, E., Tsolakidis, C., & Vratsalis, C. (2008). Virtual reality and museum: An educational application for museum education. *International Journal of Emerging Technologies in Learning (IJET)*, *3*(2008), 89–95.

CHAPTER 6

Relaxation and Well-Being

6.1 VIRTUAL REALITY FOR RELAXATION AND STRESS RELIEF

The technology development of immersive technology such as virtual reality (VR) and augmented reality (AR) opened up new avenues for human mental healthcare, relaxation, and stress relief. The technologies restricted previously to the fields of entertainment and video games are now paramount as fundamental tools in clinic practice, wellness programs, and self-stress management. The multisensory delivery potential of the environment within interactive spaces is yet another path to channel the stresses of everyday life, create emotional control, and reach psychological balance. The ability of VR and AR to provide calming environments, guide users through scripted calming behaviour, and facilitate mindful interaction positions both technologies at the forefront of mental care and technological innovation (Galkin et al., 2025; Suess & Maddock, 2025).

Why VR and AR are effective as a stress relief is intrinsic to the immersive nature of the two technologies. VR, via the head-mounted display and the sensory stimulation, immerses an individual in a completely built virtual world that can be built to soothe one and relax. AR, however, superimposes virtual items on the real world to create a compound space where the real is supplemented with visual or auditory data. The two processes facilitate temporary psychological detachment from fear sources and instill mastery and security feelings. By engaging the participant in calming virtual worlds, the brain may become dissociated from

DOI: 10.1201/9781003643395-6

pathogenic stimulation and re-associated with calming, organized stimuli (Faria et al., 2025).

Guided meditation and mindfulness via VR is probably the most studied application of VR for mental health. There are also apps such as TRIPP VR or Guided Meditation VR through which individuals can meditate in serene environments such as beaches, mountains, or forests with audio instructions and calming music (Malbos et al., 2025; Tsai et al., 2025). They not only depict worlds of nature, but cast upon them figures which are not in the world, such as floating tendrils of light or moving abstract forms, that bring wonder and distance. For anyone who can't focus while meditating in the morning, VR offers a contained, enveloping environment that commands focus and quiets distraction. It is already established in the literature that VR mindfulness training reduces physiological indicators of stress such as cortisol and heart rate, with downstream consequences for quantifiable relaxing effects (Martland et al., 2025).

Progressive muscle relaxation and directed breathing through VR is another option. By bringing together visual feedback and biofeedback sensors, experiences like MindLab and BreatheVR enable synchronizing breathing rhythms with visual feedback, i.e., opening and closing virtual objects or altering scenery. VR is immersive by nature, i.e., the user is inside the experience, removing mind drift distraction. Clinically, the interventions have been implemented among anxiety disorder patients, depression patients, or post-traumatic stress disorder patients, where the immersivity of the technology guarantees compliance and active participation as opposed to standard relaxation exercises (Birrenbach et al., 2025; Verstegen et al., 2025).

AR has also been evaluated as an add-on for stress management in the absence of VR. AR mobile apps can turn environment a warm place at the time with calming digital objects such as falling water drops, leaf fall, or sound on visual stimulation (Robbemond et al., 2025). For instance, application of AR in the healthcare sector is utilized in waiting rooms and casts calming animation on walls in such a manner that patients can forget anxiety prior to their experience with surgery. Similarly, breathing exercises communicated via AR on plain surfaces provide mobile relaxation exercises without withdrawal from the material environment. By presenting soothing stimuli in daily activities, AR improves relaxation learning in less intrusive daily living activities (Gao & Jairak, 2025).

Among the most compelling arguments in favour of VR and AR is their ability to provide restorative virtual nature experiences. Nature exposure has long been linked to psychological recovery and stress reduction but urban culture lacks exposure to nature. VR simulations of meadow, river, or forest ecosystems, for instance, in Nature Treks VR or Calm Place, offers the consumer the indulgence of experience of biophilic spaces as they are immersed in the midst of high-density urban environment. It is widely established that virtual exposure to nature is able to identify some of the restorative effects of actual contact, such as lower rates of rumination and improved mood. AR can further amplify this by the imposition of virtual vegetation, fauna, or seasonality over real city space to create hybrid spaces of the atmosphere of natural tranquility (Guille-Sanz et al., 2025; Martland et al., 2025; Pakebusch et al., 2025).

Other than relaxing participants, VR has also been applied in group therapy and collective wellness intervention. Multi-user VR environment allows several people to utilize the same virtual environment and relax or meditate simultaneously. Social bonding, as a buffer on its own against burnout and stress, is also facilitated. Wave XR and SoundSelf VR, for example, use audio-visual immersion and social bonding to cause relaxation among a group. These group interventions are most suitable in occupational well-being interventions, under which intervention for stress management is incorporated with the organizational culture. By providing a quick virtual break to the employees to relax during working hours, thus, organizations can gain mental well-being without inducing occupational stress.

Not only the ambiance of immersion space or meditation room but also the capability of such technologies to induce states of flow is the reason for the therapeutic impact of VR and AR, which reduces stress levels. Interactive immersive VR like painting with Tilt Brush or sculpting using Kingspray VR is relaxation through creativity. These tasks involve the user in creative labour without promoting high-order skill, thus enhancing self-expression and reducing stress. AR-augmented creativity apps also allow users to cast digital paintings or sculptures onto the environment, freeing joyful interaction with the environment. Interactive creativity was also shown to reduce levels of stress by distracting from exogenous sources of stress and redirecting attention towards reward-producing activity (Birrenbach et al., 2025; Riches et al., 2024).

AR and VR are even being integrated with the treatment of stress disorders. Exposure therapy, commonly used to deal with phobic fear as well as anxiety disorders, utilizes VR in exposing the patients to the stressful situation gradually within a controlled setting. Exposure is not necessarily a technical relaxation procedure, but it can perhaps decrease chronic stress as well as help in coping capacity development. Complemented by relaxation training in the same virtual environment, patients are able to learn to control their body reaction to threatening stimuli. Both of the two-tongued interventions strengthen resilience and reduce chronic stress. AR is added to complement in-situ therapeutic stimuli along the way, for example, by providing relaxation through placing patients in contact with mildly stressful real-life situations (Godoy-Gonzalez et al., 2024; Wrzeciono et al., 2024).

The number of technologies used to produce such interventions grows. Biofeedback insertion technology makes it possible for VR and AR systems to dynamically regulate the user's physiological state in real time. For instance, if a headset can read heart rate acceleration or skin conductance, virtual environment can be reduced in stress levels by shrinking animation size or dimming light. In addition to this, AR glasses have the ability to transform the surrounding environment adaptively based on levels of stress to provide adaptive and personalized stress management. The technology improves short-term experience in addition to long-term mental well-being through user education regarding internal states and control over them (Chittaro et al., 2024; Malbos et al., 2025; Zhou et al., 2024).

Accessibility of VR and AR relaxation technology is also growing. What was once only available at research centres and clinics who utilize the technology is now available using consumer-grade headsets and even mobile phones. This democratization of the technology makes the ability to take virtual relaxation therapy into one's home at relatively low cost possible. Notwithstanding the constraints, such as calling for further empirical studies of greater duration, technology-based coping strategy overdependence, and potential data privacy concerns from utilization of biofeedback sensors, the future is in sustained application of VR and AR to drive forward global mental health.

Table 6.1 presents various virtual and augmented reality relaxation interventions. There is a snapshot of the degree of intervention achievable and realistic extent of real-world application. Also, it shows the manner

TABLE 6.1 Ways of Relaxation in the Virtual World with Examples of Existing Software

Form of Relaxation in VR/AR	Description of the Approach	Examples of Existing Software
Meditation and Mindfulness	Guided virtual nature or abstract world meditation exercises, typically with voiceover and relaxing music.	TRIPP VR, Guided Meditation VR, Headspace VR
Breathing Training and Biofeedback	Pre-set breathing exercises with visual feedback; capability to track physiological indicators (e.g., heart rate) and adjust the environment based on the user state.	BreatheVR, MindLab, Unwind VR
Contact with Nature (Virtual Environments)	Sensory immersion in a forest, beach, or mountain environment for stress reduction and feeling of 'escape' from city environment.	Nature Treks VR, Calm Place, Perfect: VR Relaxation
Creative Expression	Creative pursuits in VR, i.e., virtual painting, drawing, sculpting in virtual space, unwinding through creativity.	Tilt Brush (Google), Kingspray Graffiti VR, SculptrVR
Musical and Audio-Visual relaxation	Interventions involving the use of music, visualizations, and sense interaction, generally in the form of 'sound journeys'.	SoundSelf VR, Wave XR, Melody VR
Group/Social Relaxation	Group relaxation sessions in VR rooms, generally as part of well-being programs or group therapy sessions.	Wave XR (group modes), Engage VR, AltspaceVR
AR: Enriching the Environment with Relaxing Elements	Projection of digital objects and effects onto the physical environment, e.g., leaf, water, or breathing guides animated onto walls.	AR Mindfulness App, PAUSE AR, hospital AR solutions (e.g., calming waiting room projections)

Source: Author's own work.

in which immersive technology can be employed to enable mental health through meditation, relaxation training, exposure to nature, creative activity, music and visual exposure, and social relaxation.

Table 6.2 represents only a selection of actual VR and AR applications designed to relax customers, feel more present, and de-stress. Each of the following examples illustrates how interactive or enhanced worlds can provide consumers with calming simulations, directed meditations, or virtual journeys into nature that are psychically calming.

TABLE 6.2 Real Applications of VR and AR for Relaxation and Well-Being

Application	Description
TRIPP (VR)	Mindfulness and meditation application of interactive VR environments to lead consumers through relaxation techniques, breathing, and mood-changing visual stimulation.
CalmPlace (VR)	Virtual reality application to take consumers to soothing natural environments like beaches, forests, or mountain valleys to allow tension release and enhance concentration.
Nature Treks VR	Interactive VR application in which users are allowed to navigate virtual worlds interactively, interact with wildlife, and customize environments to attain emotional release.
Healium AR/VR	Application of AR and VR relaxation exercise with biofeedback for real-time biometrics like heart rate, displaying images to relax.
FlowVR	Stress virtual reality application, meditation, and positive psychology intervention mood training and resilience.
Guided Meditation VR	Gateway to relaxation of over 100 rooms of meditation in order to foster mindfulness, breathing, and care of emotions.
Supernatural (VR)	Virtual reality exercise and relaxation system incorporating exercise with peaceful natural spaces and guided breathing in order to enable exercise and relaxation.
AR Mindfulness Apps (e.g., AR Meditation Place)	Mobile AR apps that overlay calming 3D environments and atmospheric effects over the real world to cause relaxation but not complete VR immersion.

Source: Author's own work.

6.2 AUGMENTED REALITY FOR CREATING CALMING AND STRESS-FREE SPACES

The ability to utilize AR while developing calm and serene spaces is priceless for individuals who daily are exposed to stimulation, sound, and continuous streams of information. VR places the consumer entirely inside a world of scripts, whereas AR brings computer aspects into the world as it is. This creates a mixed space where average physical spaces are supplemented by properly written digital material. One can design such an environment that maximizes comfort, increases alertness, and recovers the psychological balance with a feeling of continuous communication with the external world (Lovell & Wetherel, 2024).

The psychological rationale behind AR application in order to reduce stress stems from its ability to change sensory perception and environmental stimuli that are the most significant determinants of human welfare. Hostile environments, whether at home, work, or cities, overstimulate senses and result in excessive release of cortisol and mental exhaustion.

AR can change at the strategic level so that the sensory inputs can be anesthetized or re-directed without changing the physical world. An ordinary room, for instance, would be supplemented with computer-generated images of peaceful outdoor environments, flowing sky, or ambient nature scene like water or leaves. This selective supplementation is not intended to overwhelm reality but to build a more stabilized world of the senses that will be adaptive to relaxation. AR is the ability to seamless integration in typical situations in efforts to gain maximum psychological benefit with minimal behavioural adjustment on the part of the user (Nath et al., 2023; Singh et al., 2025).

Casual AR space design is built upon substantive borrowing from environmental psychology and biophilic design. Science has qualified human beings as having an inherent susceptibility towards such environments as are similar to evolutionary environments with abundant greenery canopy, open space, and natural light. These are linked to agreeable mood, low stress recovery time, and enhanced cognitive processing. AR introduces such biophilic factors superimposed on otherwise stressful or barren environments. A headset or wall-mounted AR device, for example, can superimpose plants, waterfalls, or birds in an office interior and thus duplicate some of the natural features that would otherwise not be present. The user is induced to feel as if the area in nature has been transformed into something cozy and aesthetically pleasing without rearranging or gaining access to the actual natural features. The emotional response elicited by these hyper-realized characteristics – composure, reduced anxiety, and enhanced sense of safety—is what attracts AR as a technology medium for facilitating human-environment interaction (Lepilkina et al., 2024).

AR can tailor its interventions according to the individual user's need and affective templates, and is therefore even more efficacious as an anti-stress intervention. With the integration of wearables monitoring biometric signals like heart rate, skin conductance, or brain waves, AR environments can measure present stress levels and react with individually adapted relaxation stimuli. For instance, upon recognition of high stress levels, the system can continuously reduce light in the augmented world, introduce calming natural sounds, or display a creeping digital effect with breathing recommendations. This responsiveness makes AR an interactive, rather than static, environment. It is dynamic, being adjusted by the subject's mood. Such feedback mechanisms are especially beneficial in therapy, where pain distraction is not just the aimed result but also the development of hardness and consolidation of self-regulation skills (Martland et al., 2025; Singh et al., 2025).

Uses of AR for work and health have already been the subject of initial research; pilot research has also occurred. AR relaxation rooms have been tried in hospital waiting rooms where patients become anxious before procedures. Casting soothing scenes, starry night skies, or waves of water has been shown to reduce measures of self-reported stress and increase patient satisfaction. In the workplace, AR experiences also are being crafted to assist in combating burnout and mental exhaustion through the delivery of micro-break experiences. Workers can, for instance, trigger an AR overlay of a peaceful beach or forest atmosphere in their office cubicle and begin short but useful relaxation sessions without ever having to leave work. These studies demonstrate the efficiency of AR in helping a varied population of users ranging from hospital patients and physicians to office workers and school kids (Birrenbach et al., 2025; Pakebusch et al., 2025; Faria et al., 2025).

Another manner in which the role contributed by AR in reducing stress is described is that it is able to share experience. While most relaxation activities are done in solitude, AR has the capability to create a shared atmosphere of relaxation across semi-public or public spaces. Urban designers and architects are designing AR installations that turn crowded city centres or transportation hubs into remapped augmented spaces where the flow of pedestrians is subjected to choreographed colours, sounds, and interactive virtual art created with the purpose of inducing relaxation. Such interventions may be able to decrease total loads of stress in more concentrated populations but increase social bonding through social experience. AR apps for applications of public mental health policy fit into broader 'smart cities' and 'digital well-being infrastructure' efforts in which technology is not necessarily being mobilized to maximize efficiency, but to improve quality of life in urban areas (Nath et al., 2023).

Designing peaceful AR environments has to deal with psychological as well as technological problems sensitively. Too much stimulation, image mis-calibration, or being too stereotypical can ironically create levels of stress instead of reducing them. In case designers do not want to take these risks, they must have their solutions grounded firmly on empirical facts regarding human perception and stress conduct. Subtlety, coherence, and user control are some of the most important principles. The extended content can never overpower or encroach on space but be an unobtrusive part of it, giving subtle hints releasing tension. User control is also a requirement; the users need to be able to control turning on, switching, or switching off the extended content as per their own volition. This is an approach that enables user control and turns AR into a less distractor technology and a more assistance technology (Gao & Jairak, 2025; Riches et al., 2024).

Technologically, the effectiveness of AR in providing stress-free environments relies on its ability to make wearable tech light in weight, ergonomically shaped, and even affordable. Existing AR headsets, however helpful, are still inconvenient to wear for a couple of hours and even likely implausible when relaxing in real life. Wearability solutions, i.e., see-through AR glasses or room-scale projection, are thus required in order to make mass diffusion of AR-relaxation a reality. Software technologies like adaptive personalization algorithms based on artificial intelligence will also increase the extent of relaxation experience that can be achieved in real-time. Utilization of AR as easy and ordinary access to mental wellness will expand exponentially with such technology implementation (Verstegen et al., 2025; Robbemond et al., 2025).

Utilization of AR to combat stress also presents some ethics. While the technology can be of hugely positive influence, the concern is becoming dependent on extended environments simply to relax. That would replace natural coping mechanisms and decrease resilience in non-augmented environments. The second concern is privacy and data security if biometric tracking is being sold with AR technology. Stress levels data, emotional state data, and physiological response data must be safeguarded so that it won't be exploited against individuals. Ethical design would then have to introduce transparency, user consent, and data protection as the central pillars in employing AR for healthcare (Nath et al., 2023; Singh et al., 2025).

AR can be incorporated into mainstream architecture and interior design. At least lighting and acoustics are already maximized for comfort, AR would become the norm for residential, office, hospital, and public building design to obtain maximum psychological well-being. 'Augmented calmness' would be a root level practice field with skills from psychology, design, technology, and urban design. The eventual outcome of this convergence would not only reduce individuals' stress levels but also strengthen societal mental health resilience against the pressures of rapid urbanization, digital overexposure, and climate change.

Table 6.3 clearly shows how AR may be implemented in different environments to bring about peaceful and relaxing environments. It highlights

TABLE 6.3 Applications of Augmented Reality for Creating Calming and Stress-Free Spaces

Context/Setting	Type of AR Intervention	Mechanism of Stress Reduction	Examples of Applications	Expected Outcomes
Healthcare (Hospitals, Clinics, Therapy Centres)	Projecting natural environments, soothing animation, or starry skies onto waiting rooms and patient rooms.	Redirecting the person away from anxiety-inducing stimuli and reduced stress.	AR projections of beach or forest landscapes in medical procedures; AR-guided meditation using AR-based technology in treatment	Reduced anxiety in patients, enhanced satisfaction, reduced perceived waiting times
Workplace (Offices, Remote Workspaces, Co-Working Hubs)	AR micro-break conditions with relaxation imagery and background soundscapes.	Replenishes thinking ability, reverses information saturation, lifts focus	Virtual forest overlay in open-office spaces; AR mindfulness training on smart glasses	Reduced risk of burnout, enhanced productivity, enhanced job satisfaction
Home Environments	Personalized AR overlays for the outdoors, artwork, or interactive light animations.	Perceptually activates the environment, provides awareness, diminishes sensory overload	AR 'Zen corner' with virtual waterfalls or interactive bonsai trees; AR starry night projections on bedrooms	Enhanced relaxation, better sleep, stress reduction in everyday life
Urban Public Spaces	Immersive AR installations with calming communal experiences (soundscapes, visual cohesion).	Stimulates but doesn't over-stimulate in high-density spots, creates social calmness, fosters sense of community	AR fountains and gardens in urban spaces; augmented murals with relaxing effects that respond to user touch	Reduced urban stress, enhanced mood in passers-by, community bonding

(Continued)

TABLE 6.3 Applications of Augmented Reality for Creating Calming and Stress-Free Spaces *(Continued)*

Context/Setting	Type of AR Intervention	Mechanism of Stress Reduction	Examples of Applications	Expected Outcomes
Educational Settings	Virtual or digitally-augmented classrooms or study rooms with natural surroundings.	Less performance anxiety, better concentration, better information retention	AR relaxation breaks between school classes; AR breathing exercises projected on classroom walls	Enhanced learning outcome, enhanced emotional regulation in students
Therapeutic and Wellness Practices	Virtual reality, or mixed.	Offers predictable relaxation cues, boosts stress-reduction practice immersion	AR headsets superimpose quiet forest trails on yoga; AR breathing exercises with visual feedback	Enhanced therapy outcome, enhanced participation, stress resilience in the long term

Source: Author's own work.

the range of environments that include workplaces and health units, home, public places, schools, and therapy centres.

6.3 POTENTIAL OF VIRTUAL AND AUGMENTED REALITY TO ENHANCE WELL-BEING

Now that workday and home stress are among the leading causes of mental states, VR and AR are likely exciting to employ for relaxation, restoration, and wellness. Both workers and dwellers in home spaces can likely benefit from properly designed immersive experiences of brief escape from daily nuisances of the day and invigorating restorative processes at mental as well as emotional levels.

In the workplace environment, worker well-being is the prime concern of human resource management, occupational psychology, and organizational development. Occupational stress is a chronic situation that leads to burnout, absenteeism, and decreased productivity, in addition to being accountable for long-term health consequences. VR and AR technologies provide varied interventions that are easily integrated into corporate wellness programs. Employees are able to relax in calming virtual settings, such as landscape, ocean, or forest, that have been scientifically associated with stress relief and parasympathetic activation of the nervous system. Slight repositioning of the VR headset can permit employees to step away from work quickly and take a sort of 'micro-holiday' that will refresh attention and increase focus. AR software goes a step further by overlaying soothing audio or visual content onto existing office spaces, creating blended environments that are less monotonous and more psychologically restorative. For example, a user of a cubicle office employs AR to project soothing waterfall on the wall or disseminates ambient sound covering open-plan office sounds. These interventions reduce stress as well as enhance creativity and emotional resilience (Gains et al., 2025; Guo et al., 2025; Wong & Leung, 2025).

In domestic contexts, VR- and AR-enabled relaxation is equally important as a means of psychological recovery. Home is usually presumed to be relaxing per se but in most situations also harbours islands of stress, ranging from inter-personal tension through home administration stress and digital clutter. Immersive technology in these cases is charged with moving one's sanctuaries spaces. VR has the potential to make a living room a meditation room, a serene garden, or a night sky in hopes of involving the users in guided relaxation, mindfulness, or simply quiet immersion in visually calming spaces. AR, for example, by casting interactive calming graphics on actual space, has the possibility of redescribing home space as therapeutic space. A humble flat is no longer confined, as AR can produce windows in virtual space or offer emerging gestural paintings responsive to gesture, reasserting agency and mastery of oneself over nearby space. They can be particularly beneficial for wooded urban rural cities, with compensatory exposure to nature-like experience with harmlessly beneficial impact on psychological health (Faure-de Baets et al., 2025; Pardini et al., 2025; Zhang & Huang, 2025).

One potential use of VR and AR for wellness is that they can be tailored. Unlike passive wellness tools such as music or static video, immersive tech can dynamically adapt to align with mental states and personal interest. In particular, brief immersion in a peaceful forest with ambient low-level noise is relieving if one worker is too mentally exhausted, while an emotionally drained worker is relieved more by interactive VR exposures evoking playfulness and pleasure. Likewise, in home use, lightness, color palette, or scene rate of the surrounded scenes can be adjusted. This adaptation reaches the highest degree of usability and productivity since the consumers themselves are involved in the process of designing their own relaxation space and not just passively consuming it. Moreover, adjustable AR and VR settings integrated with biometric sensors can monitor indicators such as heart rate variability, respiratory rate, or facial expressions and adjust the relaxation environment in real time. Such a biofeedback choice optimizes the recovery outcome and is evidence-based practice aligned for mental health treatment (Farahani et al., 2025; Gegana et al., 2025; Kim et al., 2025).

Another advantage of immersive relaxation technology is that it can democratize stress-reducing interventions. Relaxation interventions like wellness retreat, spa therapy, or ecotourism are time-consuming, cost-prohibitive, or geographically restricted. AR and VR treatments could also be employed to simulate an enormous array of such treatments for a fraction of the cost and in the setting of normal life. For employees, firms could provide communal VR terminals in corporate break rooms or compensate workers for individual headsets as an extra component of medical benefit packages. For the average customer, consumer-grade AR and VR devices available for purchase by the general public represent inexpensive access to professionally crafted wellness programs. It is a democratization that is a component of broader public health aspirations to bring down disparity in access to restorative practice and allow individuals to attain mental balance in stress states. At the same time, researchers and designers should take into account the risk of over-reliance on immersion technology so that it enhances and does not replace physical social interaction and natural attachment to physical environments (Buche et al., 2025; Reshi & Teli, 2025; Zhong et al., 2025).

Use of VR and AR for overall well-being also fits into broader theoretical models of environmental psychology and human-computer interaction. Restorative environment theory, which was originally linked to natural environments, can be transferred to digitally mediated immersive environments. Cognitive psychological studies show that experience of restorative environments recovers from fatigue of directed attention and reduces physiological indicators of stress. VR displays of forest, sea, or mountain landscapes possess much the same restoration qualities as these things: soft fascination, sense of being away, and congruence with user interest. AR's bitter-sweet experience still does suggest extension of present physical spaces without full immersion to facilitate easier integration into daily routines. These theoretical proposals more likely are aimed at immersive relaxation as not technological revolution but science-informed intervention according to dominant conceptions of human flourishing (Barrio et al., 2025).

Several considerations are to be made while using VR and AR in home and office environments. Prolonged use of immersion environments will lead to visual fatigue or motion sickness for sensitive users. Additionally, technical constraints at the cost, device weight, and accessibility level may be barriers to adoption, particularly by cheap firms or computer low-literate families. Even greater privacy concerns are at stake if biometric sensing is combined with adaptive systems. The future of AR and VR for well-being therefore requires strong ethical analysis, human-centred design, and regulation that achieves highest benefits and lowest risk. Hybrid treatment models like briefer, controlled immersive treatment with adjunctive real-life practice like mindfulness practice or physical relaxation skill training will be the most sustainable solution (Kumar et al., 2025; Oasi et al., 2025).

BIBLIOGRAPHY

Barrio, P., Andreu, M., Bruguera, P., Miquel, L., & Slater, M. (2025). Enhancing the well-being of psychiatric inpatients with substance use disorders through virtual reality-based nature exposure: A pilot study. *International Journal of Mental Health and Addiction, 2667.* https://link.springer.com/article/10.1007/s11469-025-01471-9

Birrenbach, T., Häni, S., Jegerlehner, S., Exadaktylos, A. K., & Sauter, T. C. (2025). Feasibility, subjective effectiveness, and acceptance of short virtual reality relaxation breaks for immediate perceived stress reduction in emergency physicians: Single-arm pre-post intervention study. *JMIR XR and Spatial Computing, 2*, e72605.

Buche, H., Michel, A., Anders, R., & Blanc, N. (2025). The role of engagement in virtual reality to enhance emotional well-being in breast cancer patients: A mediation analysis. *Cancers, 17*(5), 840.

Chittaro, L., Serafini, M., & Vulcano, Y. (2024). Virtual reality experiences for breathing and relaxation training: The effects of real vs. placebo biofeedback. *International Journal of Human Computer Studies, 188*, 103275.

Farahani, L. F., Mavridou, I., Zhao, J., & Klippel, A. (2025). Integrating virtual reality and physiological measures to study restorative environments for mental well-being. *Proceedings of the 2025 IEEE Conference on Virtual Reality and 3D User Interfaces Abstracts and Workshops (VRW 2025)*, 1033–1038.

Faria, S., Fonseca, S. M., Marques, A., & Queirós, C. (2025). Impact of a single virtual reality relaxation session on mental-health outcomes in frontline workers on duty during the COVID-19 pandemic: A preliminary study. *Healthcare Switzerland, 13*(12), 1434.

Faure-de Baets, J., Besnard, J., Banville, F., Cassereau, J., & Allain, P. (2025). Effects of virtual reality mindfulness on cognition and well-being in ALS: A randomized trial protocol. *Contemporary Clinical Trials, 152*, 107876.

Gains, A. M., Moghaddam, N., Baliousis, M., & Biswas, S. (2025). Virtual reality interventions for mental health and well-being in palliative care: Systematic review and narrative synthesis—Outcomes, techniques, processes and mechanisms. *BMJ Supportive and Palliative Care, 15*(5), 550–563.

Galkin, S. A., Sozonov, A. S., Bykova, M. A., Roshchina, O. V., & Bokhan, N. A. (2025). Effectiveness of using a relaxation scenario in virtual reality technology for young people. *Human Physiology, 51*(3), 245–249.

Gao, S., & Jairak, K. (2025). Virtual reality art to create a place of relaxation. *Journal of Mind and Behavior, 46*(1), 252–278.

Gegana, G. A., Limas, J. N., & Yosefanny, F. (2025). Architecturally designed virtual environment to support elderly well-being using virtual reality. In *Advances in Science, Technology and Innovation* (pp. 213–224). https://link.springer.com/chapter/10.1007/978-3-031-63061-3_19

Godoy-González, M., López-Aguilar, J., Fernández-Gonzalo, S., Jodar, M., & Navarra-Ventura, G. (2024). Efficacy and safety of a non-immersive virtual reality-based neuropsychological intervention for cognitive stimulation and relaxation in patients with critical illness: Study protocol of a randomized clinical trial (RGS-ICU). *BMC Psychiatry, 24*(1), 917.

Guillen-Sanz, H., Escolar-Llamazares, M. C., Quevedo Bayona, I., Martinez-Martin, M. A., & Bustillo, A. (2025). Can immersive virtual reality environments improve stress reduction? Experimental design with progressive muscle relaxation training. *IEEE Access, 13*, 104312–104329.

Guo, F., Li, M., Lin, G., & Zhang, D. (2025). Virtual reality tourism as a therapeutic tool: Assessing the well-being benefits of repeated restorative environment exposures for individuals with GAD. *Tourism Management, 110*, 105178.

Kim, S., Park, S., Uhm, J.-P., & Ahn, S. (2025). Exploring the social dimensions of virtual/augmented reality sport gaming: Interplay among game involvement, interpersonal relationships, social capital, and psychological well-being. *International Journal of Human Computer Interaction*. https://www.tandfonline.com/doi/full/10.1080/10447318.2025.2495113

Kumar, A., Sankhyan, S., Singh, A., Bhavsar, A., & Dutt, V. (2025). Virtual reality meets machine learning: Predicting HRV and stress outcomes for mental well-being. *Proceedings of the 2025 IEEE Conference on Virtual Reality and 3D User Interfaces Abstracts and Workshops (VRW 2025)*, 971–975.

Lepilkina, T. A., Beniashvili, A. G., Starodubova, A. V., Bogdanov, M. A., & Rodkina, S. V. (2024). Virtual reality for relaxation: Types and factors of efficiency. *Sovremennaa Terapia Psihiceskih Rasstrojstv, 2024*(1), 32–42.

Lovell, B., & Wetherell, M. A. (2024). Do virtual reality relaxation experiences alleviate stress in parents of children with autism? A pilot study. *Journal of Child and Family Studies, 33*(7), 2134–2141.

Malbos, E., Chichery, N., Borwell, B., Seimandi, J., & Lançon, C. (2025). Virtual reality and relaxation for the treatment of generalized anxiety disorder: A randomized comparative study with standard intervention. *Journal of Clinical Medicine*, *14*(4), 1351.

Martland, R., Valmaggia, L., Paleri, V., Steer, N., & Riches, S. (2025). Multi-session virtual reality relaxation for mental health staff: A feasibility and acceptability study. *Mental Health and Prevention*, *37*, 200408.

Nath, N., Kalatzis, A., & Stanley, L. (2023). Measuring user engagement in virtual, augmented, and mixed reality interventions for stress reduction. In: *Lecture notes in computer science, including subseries lecture notes in artificial intelligence and lecture notes in bioinformatics* (pp. 570–583). Springer.

Oasi, O., Rossi, C., Villani, D., & Riva, G. (2025). VR for hope: A virtual reality protocol to enhance hope, mentalization, and well-being in emerging adults. *Cyberpsychology, Behavior and Social Networking*, *28*(7), 543–545.

Pakebusch, V., Schlisio, B., Schönfisch, B., Krämer, B., & Andress, J. (2025). Virtual reality-based pain control in endometriosis: A questionnaire-based pilot study of applications for relaxation and physical activity. *Archives of Gynecology and Obstetrics*, *311*(6), 1721–1731.

Pardini, S., Calcagno, R., Genovese, A., Salvadori, E., & Ibarra, O. M. (2025). Exploring virtual reality-based reminiscence therapy on cognitive and emotional well-being in people with cognitive impairments: A scoping review. *Brain Sciences*, *15*(5), 500.

Reshi, S. N., & Teli, M. R. (2025). Exploring the dynamics of virtual reality (VR) and employee well-being: Addressing the motherhood penalty. In: *Future of HRM in a world of persistent virtual reality* (pp. 125–139). Emerald Publishing Limited.

Riches, S., Kaleva, I., Nicholson, S. L., Veling, W., & Valmaggia, L. (2024). Virtual reality relaxation for stress in young adults: A remotely delivered pilot study in participants' homes. *Journal of Technology in Behavioral Science*, *9*(4), 771–783.

Robbemond, L. M., Nijland, M., Alma, M., Veling, W., & van Driel, C. M. G. (2025). Exploring the barriers to and facilitators of using virtual reality relaxation for patients with psychiatric problems: Qualitative focus group study. *Journal of Medical Internet Research*, *27*, e65308.

Singh, B., Kaunert, C., & Dutta, P. K. (2025). Simulating mental-emotional intelligence through interactive virtual-augmented reality modules: Enhancing virtual training and stress reduction techniques for improved mental health—Mindfulness in regulating emotions. In: *Emotional intelligence in the digital era: Concepts, frameworks and applications* (pp. 29–48). Emerald Publishing Limited.

Suess, C., & Maddock, J. (2025). Understanding the influence of window views, plantscapes, and green décor in virtual reality hospital rooms on simulated acute-care patients' stress recovery and relaxation responses. *Health Environments Research and Design Journal*, *18*(3), 165–183.

Tsai, S.-H., Chao, C.-C., Lin, Y.-D., Wang, J.-H., & Liu, C. H. (2025). Immersive virtual reality cruise garden relaxation train: A novel BCI platform for depression. *IFMBE Proceedings*, *118*, 15–31.

Verstegen, A., Van Daele, T., Bonroy, B., van Loo, M., & Bernaerts, S. (2025). Designing a smartphone-based virtual reality app for relaxation: Qualitative crossover study. *JMIR Formative Research*, *9*, e62663.

Wong, I. Y. F., & Leung, L. Y. L. (2025). Explore the extended impacts on psychological well-being in older adults through application of virtual reality technology: An integrative review. *International Journal of Nursing Studies Advances*, *9*, 100371.

Wrzeciono, A., Cieślik, B., Kiper, P., Szczepańska-Gieracha, J., & Gajda, R. (2024). Non-sleep deep rest relaxation and virtual reality therapy for psychological outcomes in patients with coronary artery disease: A pilot randomized controlled trial. *Journal of Clinical Medicine*, *13*(23), 7178.

Zhang, Z.-H., & Huang, H.-C. (2025). Enhancing sustainable tourism through virtual reality: The role of collectable experiences in well-being and meaning in life. *Sustainability Switzerland*, *17*(13), 5809.

Zhong, C., Yao, L., Chen, L., Shi, L., & You, L. (2025). The use of virtual reality-assisted interventions on psychological well-being and treatment adherence among kidney transplant recipients: A randomized controlled study. *Acta Psychologica*, *253*, 104700.

Zhou, H., Chen, C., Liu, J., & Fan, C. (2024). Acute augmented effect of virtual reality (VR)—Integrated relaxation and mindfulness exercising on anxiety and insomnia symptoms: A retrospective analysis of 103 anxiety disorder patients with prominent insomnia. *Brain and Behavior*, *14*(10), e70060.

CHAPTER 7

Virtual and Augmented Reality Enhancing Modern Living

7.1 ENHANCING LIVING IN THE ANALYZED AREAS

Virtual reality (VR) and augmented reality (AR) technology is slowly transforming the way humans interact with the environment and with each other. The immersive technologies are not just entertainment gadgets, but complex systems that more and more get involved in culture, business, health, tourism, and societies. VR, completely synthetic virtual reality, and AR, the real world enriched with computer-advised information and sensory inputs, are merging to form so-called mixed reality worlds. They are enhancing human perceptual, interactional, and experiential capabilities and thus are reshaping life today. This chapter considers how VR and AR are revolutionizing life in five interlinked fields: entertainment and culture, shopping and retail, sport, tourism, and relaxation and well-being.

The first of the fields covered by this book where VR and AR have made giant strides is entertainment and culture. Cultural heritage digitization, museum experience with interactivity, and immersive theatre production are landmark uses. VR allows one to overcome spatial and temporal constraints by travelling to re-creations of ancient city centres, virtual theatres, or virtual museums from anywhere on the globe. VR offers perspectives to crowds by eliminating geography-based restrictions and allowing mobility-disabled or economically challenged individuals

DOI: 10.1201/9781003643395-7

to participate in cultural activities. AR goes a step further and overlays interactive objects onto real-world cultural activities. To illustrate, AR museum visits allow one to see artefacts in their original form, read more historical sources, or watch virtual reconstructions imposed on the actual exhibition room. In consumption of media, game and movie watching rooted in VR allow for highly personalized and emotive experiences, and AR turns passive viewing into active engagement, i.e., concerning location-based gaming phenomena. Therefore, VR and AR are shaping culture production and consumption across boundaries, revolutionizing traditional authorship, distribution, and audience participation (Dang, 2025; Irminanda & Xiong, 2025; Liu, 2025; Ramachandran et al., 2025; Zhang et al., 2025).

Immersive technology has finally revolutionized shopping and retailing. Classic linear retailing is being substituted with interactive, data-fueled, and personalized experiences. Virtual showrooms allow consumers to visualize products in three dimensions, experiment with them, and get real-time personalized advice. AR allows consumers to 'try on' clothing, accessories, or make-up virtually without ever actually touching them, thus increased buying decision confidence and fewer returns as well as logistical inefficiencies. Retail is increasingly driven by deeper customer involvement and by the ability to build stronger brand–consumer relationships through experiential marketing. Apart from this, immersive commerce also succeeds in fulfilling sustainability goals by minimizing the environmental footprint of built-in overproduction and waste transportation that is inherent in experiment-and-error consumption. The integration of VR and AR into e-commerce platforms can also respond to generational transitions in consumer behaviour by offering digitized, augmented, and game-like customer experiences that are particularly attractive to next-generation users. VR and AR are thus not just transforming the dynamics of shopping but the symbolic and experiential logics of consumption (Liu, 2025; Thandayuthapani & Thimoorthi, 2025; Elsholz et al., 2025; Grande et al., 2025; Lu et al., 2025).

VR and AR serve both functions in sports: they maximize training programs of the athlete and maximize fan experience. To the competing athlete, VR simulations offer an open, controlled, and reproducible space in which to practice. The athlete may rehearse strategy, examine biomechanical performance, and perform psychological training without physical strain or danger of injury. AR, conversely, superimposes real-time data over an exercise and thereby enables players and coaches to monitor performance parameters, heart rate, or movement patterns in real time. Regarding the fans themselves, VR is providing completely immersive experiences that recreate attending the live game, and AR products are providing additional levels of live real-time data, player statistics, and interactive content in real stadiums. The blending of VR and AR is making sport a multi-dimensional event in which performance is no less digitally mediated than watching. This aligns with the rapidly expanding e-sports ecosystem, where competitive gaming increasingly becomes a creative, technology-mediated experience that integrates physical skill, strategic decision-making, and spectacle of e-sports, competitive gaming itself is a VR and AR creative experience in which athleticism, strategy, and spectacle are being combined anew (Behlau et al., 2025; Pastel et al., 2025; Vidal-Almela et al., 2025; Wang et al., 2025).

Tourism is one of the most lively industries impacted by VR and AR. VR can permit one to visit virtually, and so pre-trip planning of journeys or even entire journeys is accessible for the immovable, the unaffordable, or the powerless. Heritage sites and natural places may be rebuilt in the virtual environment and visited by anyone with suitable hardware, rescuing threatened heritage and making global culture accessible. AR, on the other hand, enriches on-site experience through the superimposition upon reality of interpretive information. Visitors can employ AR apps, which reconstruct ruins from historic photographs, translate signs in real time, or guide cities with interactive maps. Both of these technologies lower cognitive travel barriers, open destinations to a larger market, and enable visitors' enjoyment of local heritage. Moreover, VR tourism experiences increasingly dominate education, green-propaganda campaigns, and therapy and link tourism with leisure activities and cultural diplomacy and environmentalism. Thus, in this manner, VR and AR both consolidate travel as a physical-digital mixed mode of exploration (Chan et al., 2025; Han et al., 2025; Xing et al., 2025; Guo et al., 2025; Stecuła & Naramski, 2025).

The relaxation and wellness domain reflects both therapeutic and existential fame of AR and VR technologies. Virtual worlds are utilized in psychotherapies as well, including exposure therapy of phobia, stress management training, and mindfulness training. Virtual worlds are limited and safeguarded environments where the subjects deal with fear, get conditioned to relax, or get exposed to therapeutic natural environments.

AR provides these apps to combine calming stimulation with everyday contexts, creating adapting contexts for relaxation, concentration, or therapy. Clinics and hospitals also use VR-based treatments for pain relief, motor rehabilitation, and distraction from invasive treatments. Socially, this technology allows dynamic control of health culture among citizens experiencing immersive well-being as an ordinary aspect of daily life. The potential of direct access to meditational spaces, guided treatment, or interactive cognitive therapy places VR and AR not just as therapeutic interventions but as preventive care intervention agents. Their utilization within organizational wellness programs and home health care environments reflects a growing recognition of the interconnectedness between digital technology and human health (Cieślik, 2025; Faria et al., 2025; Galkin et al., 2025; Suess & Maddock, 2025).

The above-mentioned disciplines reflect the way VR and AR, collectively, are revolutionizing life as it is. Their mutual focal point is that they both aim for the removal of spatial, temporal, and physical boundaries of human life. They make it possible to generate hybrid realities in which people can interchangeably travel from the physical reality to the VR and vice versa, documenting maximum personal realization, professional growth, and social communication. These technologies also entail urgent issues related to privacy, data protection, access, and the risk of estranging human beings from unplanned social interaction. As valuable as VR and AR are changing lives today in ways, wary integration involves wise regulation, ethical awareness, and inclusive design priorities that cannot allow for new forms of inequality or virtual dependency.

Table 7.1 implies leading fields of modern life into which virtual and augmented reality are being applied, and their comparative roles in shaping human experience. It demonstrates the way VR and AR are disruptive technologies cutting across entertainment and culture, retail and shopping, sport, travel, and relaxation and well-being. In each of these fields, VR tends to create fully virtual, immersive worlds, whereas AR extends pre-existing real spaces by adding interactive and context-dependent overlaying. In collaboration, they maximize access to cultural heritage, personalize consumption, optimize sport training and viewing, enable distant and augmented travel experience, and provide preventative and therapeutic healthcare. The combined impact on industries and lifestyles is so monumental. The table thus sketches the variety of VR/AR applications and how they will most likely change day-to-day life in industries.

TABLE 7.1 Domains of Modern Living and the Role of VR/AR

Domain	Role of VR/AR
Entertainment and Culture	VR constructs simulate cultures like virtual music concerts, drama, and museums; AR enriches existing exhibits with interactive and didactic overlays.
Shopping and Retail	VR offers virtual showrooms and product demonstrations; AR offers virtual try-on of clothing, accessories, and make-up and augmented personalized shopping.
Sport	VR provides safe training simulation and tactical rehearsals; AR provides real-time projection of data and analytics onto training and live events for players and audiences.
Tourism	VR offers virtualized access to locations and computer-simulated models of past locations; AR provides on-location interaction with related information and support.
Relaxation and Well-Being	VR offers restorative and soothing settings to reduce stress, rehabilitate, and distract from pain; AR introduces adaptive calming stimuli into daily environments.

Source: Author's own work.

7.2 GENERAL BENEFITS OF USING VIRTUAL AND AUGMENTED REALITY

Table 7.2 is an overview of how VR and AR can be useful in day-to-day life. Nine most significant areas – culture and entertainment, education and training, shopping and retailing, sport and exercise, tourism, health and well-being, workplace and professional life, social interaction, and research and innovation – are listed in the table and an overview of how each of these is enabled by VR and AR technologies. Although both

TABLE 7.2 Benefits of Using Virtual and Augmented Reality

Domain	Benefits of Virtual Reality	Benefits of Augmented Reality
Entertainment and Culture	Enables completely immersive gaming, cinematic, concert, and theatrical experiences; allows global publics to virtually visit virtual exhibitions and reconstructed historic sites; allows access for the mobility- or cost-imporadential; encourages creativity and new culture.	Enhances physical performances and deliveries with interactive features; allows real-time visualization of historic buildings; allows multilingual capability; allows interactive storytelling and audience interaction.

(Continued)

TABLE 7.2 Benefits of Using Virtual and Augmented Reality *(Continued)*

Domain	Benefits of Virtual Reality	Benefits of Augmented Reality
Education and Training	Offers realistic, risk-free simulation for medicine, aviation, engineering, and the military; lowers training costs; enhances participation and retention; allows distance learning through collaborative virtual classrooms; allows collaboration worldwide across disciplines.	Allows real-time technical tips and advice overlays; allows learning through understanding in context; allows interactive textbooks and laboratory sessions; allows 3D models to be interactively created for the students to learn classwise; enhances greater levels of engagement in STEM learning.
Shopping and Retail	Facilitates experiential 3D showrooms and virtual shopping malls; facilitates discovery of products in imitative environments (e.g., furniture in a living environment); builds confidence in purchasing; minimizes returns; facilitates sustainable consumption by restraining excess production.	Enables online fitting of garments, footwear, and makeup; enables product visualization in real-world settings before buying; provides interactive offers; enhances shopping in stores with digital overlay; accelerates decision making.
Sport and Fitness	Facilitates replicable and controlled training environments; optimizes tactical and psychological readiness; provides biomechanical and cognitive analysis; facilitates fans to feel immersive participation in live gaming or training; facilitates growth of e-sports ecosystems.	Displays real-time performance statistics, heart rate, and tactical overlays; enhances stadium experience with stats and visualizations; enables augmented training sessions for sportspeople; provides fans with interactive real-time facts.
Tourism	Enables distant visits to locations at home; reconstructs deteriorating or destroyed sites in a virtual form for access worldwide; supports travel planning with realistic previews; enables accessible tourism for the disabled or low-income.	Complements in-situ experience with re-enactment of historical events, translation, and context-specific advice; offers city tour using AR maps; allows environmental education with real-time environmental information; allows adaptive travel timing for tourists.

(Continued)

TABLE 7.2 Benefits of Using Virtual and Augmented Reality *(Continued)*

Domain	Benefits of Virtual Reality	Benefits of Augmented Reality
Health and Well-Being	Supports phobia, PTSD, and anxiety therapy through exposure; enables relaxing environments for mindfulness and stress relief; diverts patients from pain with treatments; supports motor rehabilitation; encourages patients with gamified therapy.	Offers calming or exciting stimuli to the regular environment; allows rehabilitation with real-time movement feedback; allows disabled access; allows hospital navigation; incorporates relaxation content into regular activities.
Workplace and Professional Life	Enables global teamwork using virtual offices, conferencing, and workshops; saves cost of business travel and carbon emissions; enables 3D immersive prototyping and designing; enables virtual recruitment simulation; enables better employee training programs.	Maps virtual information to physical workspaces to guide activities; supports interactive presentations; guides employees through step-by-step directions in technical workflows; extends collaboration by integrating physical and virtual data in real time.
Social Interaction	Enables shared immersive spaces for global gatherings, festivals, and communities; enables users to 'be present' regardless of distance; saves from social isolation; enables expression of identity in virtual societies; enables inclusion in digital societies.	Improves face-to-face meetings with shared digital data; supports hybrid conferences and events; supports collaborative design and brainstorming; supports interactive group work in education, workplace, and play settings.
Research and Innovation	Offers virtual environments for safe scientific experimentation; streamlines prototyping loops; enables immersive data visualization for sophisticated data analysis; enables interdisciplinary cross-disciplinary studies between groups; enables citizen science using communal VR labs.	Complements field and lab observations with context overlays; incorporates technical task precision with electronic prompting; provides for real-time annotation of experiments; supports research training through enhanced demonstrations.

Source: Author's own work.

technologies try to extend human longevity in the ways of more senses, cognition, and social interaction, they are different in application: VR places the consumer into a complete virtual world, while AR puts digital information on top of reality. Both, in parallel co-existence, form a powerful double impulse to drive quality of life, reduce the threshold for entry, and spur innovation.

The entertainment and cultural industry unleashes the quintessential potential of immersion technology to revolutionize human consumption and creativity. VR places the consumer in a position to view concerts, theatre, cinema, and video games in highly immersive settings that eliminate geospatial, temporal, and financial limitations. Consumers are in a position to visit virtual museums and art galleries, converse with reconstructed heritage sites, and experience masterpieces on novel planes. More significantly, this access democratization is based on cultural accessibility that enables individuals with disabilities or limited resources to be admitted to world cultural life. AR makes it possible because it increases physical presentations through superimpositional interactability. Audience members, for instance, can access AR programs that allow them to see historic reconstructions of remains, receive multilingual explanation, or have interactive narration for live performance. The convergence of VR and AR hence redesigns cultural interaction: VR goes beyond physical space to cultural encounter, and AR augments distant communication with more meaning (Al Balushi et al., 2024; Gilberto et al., 2024; Kattakinda et al., 2024; Meadows, 2023; Morales-Rosario et al., 2024).

VR and AR in training and education reveal a new congruence. VR works best in high-precision, high-consequence medicine, aviation, and engineering fields where the simulations produce a realistic environment in which students can practice procedures in reproducible, controlled conditions. It is cost- and risk-saving and enhances memory retention for gains with experiential learning. Virtual classrooms that can handle multiple users also facilitate distance learning and worldwide collaboration and are transforming the structure of education in boundaryless forms. AR, on the other hand, places instruction into real space and real time by overlaying technical instruction and situation data onto tasks. Most ideally suited for vocational training, AR can lead apprentices in executing accurate manual steps, or in the classroom, where 3D AR models illuminate abstractions. When combined, AR and VR extend the classroom far beyond its conventional limits by enabling safe simulated experiences and enriching learning within real-world environments (Stecuła, 2023a; Stecuła, 2023b; Stecuła, 2025).

The retailing industry demonstrates business potential for immersive technology. VR allows one to construct fully virtual showrooms, virtual malls, and interactive spaces in which customers can inspect the products in detail and imagine them in some other context, e.g., a piece of furniture in a living room. This interaction decreases uncertainty, stimulates consumer confidence, and supports sustainable consumption by minimizing unwanted returns. AR transfers all these advantages to daily shopping in the guise of virtual trying on of clothes, shoes, or makeup, or seeing products in use prior to buying them. For companies, AR offers experience marketing depth in terms of merging the thrill of digital interaction with actual touch with products. Complementarity of AR and VR renders the retail transactional and experiential too and allows new levels of customer and brand loyalty (Elsholz et al., 2025; Rosman et al., 2025; Stecuła, 2024; Vona et al., 2025; Wolniak & Stecuła, 2024).

VR and AR react to fitness and sports with viewing and performance intention. VR offers the game players virtual spaces where practice can be replicated safely, tactics, psychological, and biomechanics. Immersion is also offered to the spectator, and they can see real game environments conveniently from home ease. AR operates in real time in real training or real events. Players and coaches can also see real-time overviews of strategic tips, health grades, or performance data during practice time, while the fans in the stadium seats get live stats, player information, and interactivity. It is this symbiotic realm that praises the way VR compresses simulation and immersion into a mere proposition, as AR compresses real-time augmentation into a like proposition, to allow for an inclusive platform in sporting development and spectatorship (Li et al., 2025; Martinez-Cevallos et al., 2025; Tang et al., 2025; Toh et al., 2025; Yue et al., 2025).

Travel is also one sector where VR and AR together transform. VR provides simulated tours of nature sites, cultural sites, and cities and makes them accessible to individuals who cannot move, are ill, or politically cannot move. VR also promotes ecotourism as it provides simulated tours of vulnerable or endangered sites, which reduce the physical loads on most tourists. AR complements ground touring with context-aware layering, real-time translation, adaptive maps, and interactive guides. Travellers can be viewing historical restorations, collecting ecological information in nature parks, or being provided with customized excursions in line with their interests. In tandem, VR and AR redefine travel as a hybrid tourism experience, and travelling no longer completely physical but with the digital space augmented, so travellers can be more immersed and have more access.

The health and wellness sector emphasizes the humanitarian side of immersive technologies. Already, VR has been found to be effective in a treatment context for diseases like phobia therapy, post-traumatic stress disorder, and anxiety disorders. Offering calming natural environments for stress reduction and mindfulness is another role played by VR. It also gamifies rehabilitation therapy in rehabilitated patients. VR is also being used more and more these days in hospitals for distraction from pain during surgery. AR enables well-being by locating adaptive calming stimuli within the structure of daily life, enabling rehabilitation through real-time feedback from movement, and widening reach in health settings through way-finding. The technologies' complementary function is revealed: VR constructs robust therapeutic settings, and AR weaves support for well-being within the structure of daily life (Malbos et al., 2025; Martland et al., 2025; Pakebusch et al., 2025).

VR and AR are transforming organizational transformation at work and in business. VR eliminates the costs of travel and carbon footprints using virtual offices, webinars, and meetings. VR facilitates experience prototyping and design procedures that allow teams to prototype products virtually prior to production. VR simulations that provide employees with genuine, performance-based feedback also facilitate onboarding and training employees. AR, however, then bridges virtual learning with actual workplaces, guides employees step-by-step in tasks, enriches interactive presentations, and facilitates easy collaboration with shared physical-digital tools. The application of VR and AR thereby makes the company more sustainable, efficient, and inclusive and enriches creativity and innovation (Stecuła, 2023b; Wieczorek & Stecuła, 2025).

Social interaction as a discipline is interested in the way virtual and augmented reality redefine human relationships. VR space constructs entirely virtual worlds for international meetups, parties, and world-construction that allow participants to join regardless of where they are or have mobility impairments. This dis-isolates and opens new terrain of identity formation and virtual belonging. AR expands face-to-face interactions by overlaying co-authored digital information over body-to-body interaction to enhance hybrid conferences, co-designing, and co-learning. AR and

VR join forces to bring social interaction into hybrid spaces, where real presence and virtual presence co-exist with one another in an attempt to promote human-to-human communication and co-operation (Birrenbach et al., 2025; Stecuła, 2024; Tsai et al., 2025).

Science is the discipline where AR and VR show how they advance science. Virtual worlds are provided by VR for testing, minimize prototyping time, and enable visualization immersion of big data sets. This promotes inter-disciplinary collaboration, i.e., design thinking and engineering application. AR facilitates field and laboratory investigation using context-specific overtones, real-time annotation, and concurrent commentary that improve reproducibility and results accuracy. Combining the two technologies offers exciting opportunities for the simplification of discovery in many fields of scholarship, citizen science, and international research collaboration.

7.3 VIRTUAL AND AUGMENTED REALITY IN THE CONTEXT OF SMART CITIES

Smart city is the urban development paradigm of our time, which combines digital technology, sustainability, and participatory governance in a strategy that claims people's welfare as much as that of sound and effective urban systems. Along the way, VR and AR are also critical as decision-enabling technologies to recapture experience, governance, and planning of cities. Their role ranges from civic participation and intercultural communication to effective management of infrastructure, transportation, urban utilities, and environment for health. With the use of immersive and interactive layers within the urban context, VR and AR make the city not only effective but also participatory, inclusive, and responsive to the needs of diverse populations.

Urban planning and design are the first usage of VR and AR in smart city functions. Planners previously relied on two-dimensional models, maps, and static simulations. With VR, there is a more immersive option in the sense of allowing planners, citizens, and architects to walk through cities physically at the planning stage prior to construction. This allows stakeholders to experience spatial relationships, calculate environmental impacts, and solve conflicts in real time. They can learn about participatory planning through direct experience with proposed modeled buildings and offer opinions about things from experience instead of theory. AR goes one step further and puts proposed development over the actual environment itself so that the viewer can actually view how a proposed park, bus stop, or apartment building would change their surroundings. Such digital twin integration makes city planning an interactive, dynamic, and democratic process by default for experts and for the citizens (Kozhevnikov & Svitek, 2025; Wajid et al., 2025; Wiltenburg et al., 2025).

In addition to planning, VR and AR are a component of every urban mobility network. Smart cities' ace up the sleeve is smart transport systems and redefine efficiency as sustainability, and virtual/augmented/mixed reality make planning and customer experience child's play. VR models traffic flows, emulates autonomous vehicle conditions, and certifies operators or drivers in the urban environment. Simulations reduce risk and cost but allow novel mobility concepts such as fleet-shared self-driving vehicles or dynamic traffic management to be piloted. AR can provide real-time information to the average tourist, overlaying directions, safety notices, or accessibility information on to the real world. AR security systems can assist pedestrians and cyclists, and public transport users can be provided with real-time context-aware information about delay, timetabling, or services close by. Thus, VR ensures systematic city planning and AR daily city transport experience which drives the whole smart city goal of efficient, green, and comfortable transport (Cenkner et al., 2025; del Campo et al., 2024; Jonek-Kowalska & Wolniak, 2025; Li, 2024; Rosak-Szyrocka & Wolniak, 2025).

The virtual and augmented realities also enable the social and cultural dimension of smart cities. Virtual city culture is no longer about body space and institutional places like museums, the theatre, or concert halls but about intermedia spaces of culture where VR and AR hold the key to access, interaction, and production. And by means of VR, citizens can have access to virtual exhibitions, concerts, or community events regardless of geographical location and thereby democratize participation and access to culture. AR, on its part, infuses physical cultural spaces with an interactive story, reconstructions, and interpretive content in the very walls of the built environment itself. This dual functionality thrusts collective memory and cultural identity onto a focal pedestal, placing cities centre stage in innovation and creativity. Dear above all, perhaps, is that egalitarian nature of this enrichment in so far as it allows excluded citizens to partake in the access to civic life and cultural experience otherwise far beyond their reach. In smart cities, therefore, VR and AR are less technological innovations per se than tools of culture and citizenship democratization (Jonek-Kowalska & Wolniak, 2021; Mahdi et al., 2024; Michalik et al., 2022; Skotnicka-Zasadzień & Wolniak, 2026; Sultana & Kumar, 2024; Wolniak & Stecuła, 2024).

Health and well-being is another area where VR and AR board the smart city bandwagon. Today's cities are planning to deliver quality and equal health care, and to this, immersive technologies now promise something new. VR is being applied in the therapy of mental illness, i.e., phobia, anxiety, and post-traumatic stress disorder treatment with exposure therapy. VR is being employed for pain relief, rehabilitation, and mindfulness training, which can potentially be integrated into smart city health care systems. AR is supplementing these efforts with real-time medical data overlay to patients and doctors, rehabilitation exercise learning, and general wellness through adaptive digital stimulation in life. AR and VR are used in smart hospitals for diagnosis, for surgical training, and for teaching. AR-facilitated smart homes can aid ageing citizens in the monitoring of health parameters and adaptive feedback. With VR and AR in urban health centres, smart cities make their cities more humane, equitable, and resilient (Kohli et al., 2022; Parikh et al., 2022; Zhang, 2021).

Lifelong learning and education are a natural aspect of urban life that assumes another intensity if considered from the angle of VR and AR. The intelligent city or the knowledge city is founded on lifelong learning by citizens, and VR is the way forward to getting there. VR allows us to experience virtual classrooms, virtual laboratories, and world learning collaboration, whereby students in a learning room work together and also with other students and teachers from the same virtual learning room. This does away with physical locations and allows for cross-border knowledge exchange. AR facilitates in-situ learning, providing context-based just-in-time information to pupils in class, workers in the firm, and citizens on the street. AR supposedly can guide learners step by step through complex manual procedures and thus bridge the theory-practice gap in technical and vocational education. Given that the city is leaning towards knowledge-based economies, VR and AR help in empowering the citizenry with digital age skills that drive human capital and education inclusivity (Marques et al., 2025; Nasralla, 2021).

Urban governance and civic participation should also be taken into consideration. A smart city is not only technological but political as well since it reconfigures the ways of interaction between government institutions and citizens. VR can offer real-time interactive simulation of policy choices so that citizens can experience the impact of city policies like zoning, environmental policy, or investing in infrastructure. VR offers participatory democracy since citizens are able to provide inputs to decision-making on the basis of well-informed views. AR also offers open government since it offers relevant information regarding city services, spending, or air and water quality in the surroundings of the citizens. Humans can, for example, sense air quality from AR smartphone applications or the energy efficiency of surrounding buildings in real time. Such a product facilitates trust, transparency, and citizen engagement to make the smart city government more legitimate (Li, 2024; Mahdi et al., 2024; Vasudev et al., 2025).

Economically, AR and VR are now developing new business ecosystems in smart cities. They enable distant business, tourism, entertainment, and professional services and generate employment and innovation. Entrepreneur and start-ups use VR for virtual product design, virtual stores, or solution design, and AR services in customer and store experience. Preview destination experience prior to visit by VR makes planning and marketing easier, and AR makes navigation and interpretation easier at the site. They form hybrid economic spaces where material and virtual value creation are intertwined. AR and VR, for example, not only advance urban quality of life but also introduce economic rivalry and entrepreneurship in global smart cities' smart city networks. (Kaliappan & Marannan, 2023; Alzahrani & Alfouzan, 2022; Skubis et al., 2024; Wolniak & Stecuła, 2024). AR and VR enable powerful and resilient smart cities.

When the cities of the climate change, depletion of natural resources, and social inequality cluster experience crisis, immersive technologies bring cities flexibility and ability to plan. VR simulation assists the cities to simulate disaster, experiment with climate change adaptation, and practice exercising emergency response systems. It reduces risk and increases readiness in the context of increased environmental uncertainty. AR, however, provides the citizens with real-time environmental data, used in an attempt to catalyse sustainability practices such as energy efficiency, waste reduction, and ecogreen transport. In the meantime, the technologies put smart city on the trajectory of sustainable development where digital technology combines ecological awareness and social justice (Cenkner et al., 2025; Jonek-Kowalska & Wolniak, 2022; Jonek-Kowalska & Wolniak, 2023; Wajid et al., 2025). Table 7.3 illustrates how VR and AR enhance various aspects of city life from planning and mobility to culture, health, education, governance, business, and sustainability under the smart city canopy.

The main role of VR is to establish totally realistic virtual worlds like simulation of infrastructures built, virtual classrooms, or virtual bazaars and provides the ability to explore, co-design, and access anywhere on this

TABLE 7.3 Applications and Benefits of VR and AR in Smart Cities

Domain	Role of Virtual Reality	Role of Augmented Reality
Urban Planning and Design	Enables realistic simulation of future infrastructure; allows planners and citizens to 'walk through' virtual models of future works; allows participatory planning.	Superimposes plans over the actual world; allows residents to see forward-planned parks, roads, or buildings on the ground; makes planning more transparent.
Mobility and Transportation	Models traffic levels and behaviour for driverless vehicles; trains drivers and operators safely and in control; trials new systems of mobility.	Shows live navigation, warning for safety, and access information; overlays traffic and transport data for pedestrians, cyclists, and motorists.
Culture and Social Life	Builds immersive museums, theatres, and concerts; simulates heritage places for global public use; allows cultural inclusiveness.	Supplements exhibitions with interactive digital overlays; provides multilingual features and contextual storytelling; superimposes AR effects on live events.
Healthcare and Well-Being	Enables therapy (stress, PTSD, phobia), pain management, and rehabilitation; creates soothing spaces for mindfulness and patient care.	Provides health tips and rehab insights in real-time; puts health information into intelligent hospitals and homes; enables everyday well-being.
Education and Lifelong Learning	Builds virtual classrooms and labs; enables virtual classroom access everywhere in the world; enables learning retention through experiential learning.	Does real-time contextual information; enables technical and vocational schooling on the basis of AR overlays; enables comprehension on the basis of interactive models.
Governance and Participation	Replicates citizens' impact on policy as an experiment-before-you-try process; enables participatory democracy.	Puts information about budgets, air quality, or city services on top of physical space; makes government transparent.
Economy and Commerce	Builds virtual shopping malls, showrooms, and immersive business services; enables creative industries and start-ups.	Enables interactive shopping with virtual try-on and store AR apps; engages more engaged consumers with data-driven personalization.
Sustainability and Resilience	Recreates disaster situations and adaptation measures for climate change in the virtual environment; enables disaster training.	Provides real-time environmental feedback (air quality, energy usage, waste); encourages sustainable behaviour through contextual overlays.

Source: Author's own work.

planet. AR, in contrast, enhances reality when live overlays, real-time contextual info, and interactive guidance support decision-making, daily life, and openness. The technologies are complementary technologies combined that can make cities smarter, inclusive, efficient, and resilient when they are in close alignment to engaging citizens, city systems, and digital infrastructure.

BIBLIOGRAPHY

Al Balushi, J. S. G., Al Jabri, M. I. A., Palarimath, S., Radhakrishnan, V., & Thumu, M. B. (2024). Virtual and augmented reality in interactive entertainment: Unlocking new dimensions of immersive interaction. *Proceedings of the 2024 2nd International Conference on Computing and Data Analytics (ICCDA 2024).*

Alzahrani, N. M., & Alfouzan, F. A. (2022). Augmented reality (AR) and cybersecurity for smart cities—A systematic literature review. *Sensors, 22*(7), 2792.

Behlau, C., Pauly, H., Dreiskaemper, D., & Strauss, B. (2025). Are you thinking what I am thinking? A feasibility study of a virtual reality measurement for shared mental models in team sports. *Team Performance Management, 31*(3–4), 145–169.

Birrenbach, T., Häni, S., Jegerlehner, S., Exadaktylos, A. K., & Sauter, T. C. (2025). Feasibility, subjective effectiveness, and acceptance of short virtual reality relaxation breaks for immediate perceived stress reduction in emergency physicians: Single-arm pre-post intervention study. *JMIR XR and Spatial Computing, 2*, e72605.

Cenkner, M., Michel-Schneider, U., Tylová, N., & Bouchner, P. (2025). Virtual reality as a training tool for utility vehicle operators in smart cities. *Transactions on Transport Sciences, 16*, 49–53.

Chan, C.-S., Wong, S. Y., Agapito, D., & Tam, V. (2025). The effect of pre-trip virtual reality and on-site smart device use on accessible tourism experiences. *Sustainable Futures, 10*, 100811.

Cieślik, B. (2025). Mechanisms of virtual reality-based relaxation in older adults: A scoping review. *Journal of Clinical Medicine, 14(*17*)*, 6126.

Dang, J. (2025). Design of political classroom interaction system in social virtual reality entertainment environment based on convolutional neural network. *Entertainment Computing, 52*, 100860.

del Campo, G., Saavedra, E., Piovano, L., Luque, F., & Santamaria, A. (2024). Virtual reality and internet of things based digital twin for smart City cross-domain interoperability. *Applied Sciences, 14(7)*, 2747.

Elsholz, S., Korbel, J. J., Pham, K., & Zarnekow, R. (2025). From immersive worlds to virtual showrooms: A taxonomy of virtual reality shopping applications. *Frontiers in Virtual Reality, 6*, 1558475.

Elsholz, S., Korbel, J. J., & Zarnekow, R. (2025). Shopping in a virtual reality: A stimulus-organism-response (s-o-r) based systematic literature review. In: *Lecture notes in information systems and organisation* (pp. 381–397). Springer.

Faria, S., Fonseca, S. M., Marques, A., & Queirós, C. (2025). Impact of a single virtual reality relaxation session on mental-health outcomes in frontline workers on duty during the COVID-19 pandemic: A preliminary study. *Healthcare*, *13*(12), 1434.

Galkin, S. A., Sozonov, A. S., Bykova, M. A., Roshchina, O. V., & Bokhan, N. A. (2025). Effectiveness of using a relaxation scenario in virtual reality technology for young people. *Human Physiology*, *51*(3), 245–249.

Gilberto, L. G., Bermejo, F. R., Tommasini, F. C., & García Bauza, C. (2024). Virtual reality audio game for entertainment and sound localization training. *ACM Transactions on Applied Perception*, *22*(1), 4.

Grande, R., Albusac, J., Herrera, V., Vallejo, D., & Castro-Schez, J. J. (2025). Enhancing hand interactions and accessibility in virtual reality environments for users with motor disabilities: A practical case study on VR-shopping. *IEEE Access*, *13*, 44062–44089.

Guo, F., Li, M., Lin, G., & Zhang, D. (2025). Virtual reality tourism as a therapeutic tool: Assessing the well-being benefits of repeated restorative environment exposures for individuals with GAD. *Tourism Management*, *110*, 105178.

Han, B., Yu, J., & Bi, M. (2025). A rural tourism promotion mechanism based on virtual reality technology for real-time interactive experience. *Systems and Soft Computing*, *7*, 200299.

Irminanda, K. R., & Xiong, S. (2025). Ergonomic risks in immersive virtual reality gaming for entertainment: A literature synthesis. *International Journal of Human Computer Interaction*. https://doi.org/10.1080/10447318.2025.2490402

Jonek-Kowalska, I., & Wolniak, R. (2021). Economic opportunities for creating smart cities in Poland: Does wealth matter? *Cities*, *114*, 103222.

Jonek-Kowalska, I., & Wolniak, R. (2022). Sharing economies' initiatives in municipal authorities' perspective: Research evidence from Poland in the context of smart cities' development. *Sustainability*, *14*, 4.

Jonek-Kowalska, I., & Wolniak, R. (2023). *Smart cities in Poland: Towards sustainability and a better quality of life?* Routledge.

Jonek-Kowalska, I., & Wolniak, R. (2025). *Exclusion in smart cities: Assessment and strategy for strengthening resilience*. Routledge.

Kaliappan, S., & Maranan, R. (2023). Implementation of cyber physical systems in smart cities through augmented reality networks in the mobility decade. In: *Cyber physical systems and supporting technologies for industrial automation* (pp. 215–228). IGI Global.

Kattakinda, J. P., Dhananjaya, S., Mojžiš, S., Apsan, R., & Malavolta, I. (2024). Towards understanding the energy consumption of virtual reality applications in gaming, education, and entertainment. *Proceedings of the 2024 IEEE/ACM 8th International Workshop on Games and Software Engineering (GAS 2024)*, 50–57.

Kohli, V., Tripathi, U., Chamola, V., Rout, B. K., & Kanhere, S. S. (2022). A review on virtual reality and augmented reality use-cases of brain computer interface based applications for smart cities. *Microprocessors and Microsystems*, *88*, 104392.

Kozhevnikov, S., & Svitek, M. (2025). Integrating challenge-based learning and virtual reality representation in smart city education program. *Proceedings of the 12th Euro American Conference on Telematics and Information Systems (EATIS 2024)*, 18.

Li, Y.-H., Lin, C.-S., Chuang, C.-J., Hsu, J.-L., & Li, Y.-J. (2025). Applying MSEM to analyze people's cognitive behavior towards virtual reality sport experience. *Computer Science and Information Systems, 22*(3), 1361–1378.

Li, Z. (2024). Virtual and augmented reality-based environmental pollution analysis in smart city using wireless sensor network enabled hypertext system. *Computer Aided Design and Applications, 21*(S17), 250–269.

Liu, J. (2025). Development of interactive English e-learning video entertainment teaching environment based on virtual reality and game teaching emotion analysis. *Entertainment Computing, 52*, 100884.

Liu, T., Yu, W., Huang, L., Wu, J., & Pelowski, M. (2025). Collaborative online shopping with virtual reality: An explorative investigation using fNIRS-based hyperscanning. *Information and Management, 62*(8), 104218.

Lu, Y., You, W., Zheng, Z., Yang, C., & Zhou, Z. (2025). PaRUS: A virtual reality shopping method focusing on contextual information between products and real usage scenes. *IEEE Transactions on Visualization and Computer Graphics, 31*(5), 2684–2693.

Mahdi, H. H. J., Fouad, L., Hussain, F. H. T., Mohammed, M. A., & Othman, N. A. (2024). Augmented and virtual reality services supported by 6G for improving smart cities. *ICETAS 2024: 9th IEEE International Conference on Engineering Technologies and Applied Sciences.*

Malbos, E., Chichery, N., Borwell, B., Seimandi, J., & Lançon, C. (2025). Virtual reality and relaxation for the treatment of generalized anxiety disorder: A randomized comparative study with standard intervention. *Journal of Clinical Medicine, 14*(4), 1351.

Marques, M. M., Ferreira-Santos, J., Rodrigues, R., & Pombo, L. (2025). Mobile augmented reality games towards smart learning city environments: Learning about sustainability. *Computers, 14*(7), 267.

Martinez-Cevallos, D., Jimenez, M. A., Campos, C. P., & Calabuig, F. (2025). Beyond reality: The role of innovation in virtual sport events. *Engineering Economics, 36*(2), 226–240.

Martland, R., Valmaggia, L., Paleri, V., Steer, N., & Riches, S. (2025). Multi-session virtual reality relaxation for mental health staff: A feasibility and acceptability study. *Mental Health and Prevention, 37*, 200408.

Meadows, C. W. (2023). Entertainment or engagement? An examination of transportation and gender in virtual reality. *Sage Open, 13*(3). https://journals.sagepub.com/doi/10.1177/21582440231192953?utm_source=researchgate.net&utm_medium=article

Michalik, D., Kohl, P., & Kummert, A. (2022). Smart cities and innovations: Addressing user acceptance with virtual reality and digital twin city. *IET Smart Cities, 4*(4), 292–307.

Morales-Rosario, M., Tierra-Llanga, A., Morales-Gordon, J., Hernandez-Ambato, J., & Isa-Jara, R. (2024). An immersive system for light vehicles in entertainment apps using virtual reality. *ETCM 2024: 8th Ecuador Technical Chapters Meeting.* https://ieeexplore.ieee.org/document/10746074

Nasralla, M. M. (2021). Sustainable virtual reality patient rehabilitation systems with IoT sensors using virtual smart cities. *Sustainability, 13*(9), 4716.

Nikkitha, B., Chinmayie, V. B. S., Divya Shakthi, R., & Jaison, B. (2025). Tryly: Adaptive virtual fitting room and personal assistant—Enhancing online shopping with augmented reality and machine learning. *Proceedings of the 2nd International Conference on Research Methodologies in Knowledge Management, Artificial Intelligence and Telecommunication Engineering (RMKMATE 2025).*

Pakebusch, V., Schlisio, B., Schönfisch, B., Krämer, B., & Andress, J. (2025). Virtual reality-based pain control in endometriosis: A questionnaire-based pilot study of applications for relaxation and physical activity. *Archives of Gynecology and Obstetrics, 311*(6), 1721–1731.

Parikh, T., Egendorf, S. P., Murray, I., Smiley, K., & Kao-Kniffin, J. (2022). Greening the virtual smart city: Accelerating peer-to-peer learning in urban agriculture with virtual reality environments. *Frontiers in Sustainable Cities, 3*, 815937.

Pastel, S., Klenk, F., Bürger, D., Heilmann, F., & Witte, K. (2025). Reliability and validity of a self-developed virtual reality-based test battery for assessing motor skills in sports performance. *Scientific Reports, 15*(1), 6256.

Ramachandran, K. K., Lakshmi, K. K., Raja, M., Raj, M. G., & Ziora, L. (2025). Using 5G for virtual reality applications in entertainment and gaming. *Recent Trends in Engineering and Science for Resource Optimization and Sustainable Development*, 53–57.

Rosak-Szyrocka, J., & Wolniak, R. (2025). *Smart universities in smart cities: Shaping the future of education and urban innovation.* Routledge.

Rosman, D. H., Albakri, I. F. A. M., Mokhtar, M. K., Nor'a, M. N. A., & Zulkernain, M. Z. (2025). Virtual emporium: Elevating shopping experience with virtual reality in digital mall. *IFMBE Proceedings, 128*, 747–759.

Singh, B., Kaunert, C., Vig, K., & Gautam, R. (2024). Revolutionizing customers shopping experience via augmented-virtual reality in retail business: Whirling dreams toward transforming retail industry reality. In: *Augmenting retail reality Part B: Blockchain, AR, VR and AI* (pp. 23–37). Emerald Publishing Limited.

Skotnicka-Zasadzień, B., & Wolniak, R. (2026). Artificial intelligence in public administration. In: *Routledge studies in innovation, organizations and technology.* Routledge.

Skubis, I., Wolniak, R., & Grebski, W. W. (2024). AI and human-centric approach in smart cities management: Case studies from Silesian and lesser Poland voivodships. *Sustainability, 16*(18), 8279.

Stecuła, K. (2023a). Identifying diverse uses of virtual reality in higher education and exploring perceptions of VR in the chosen field. *Zeszyty Naukowe Politechniki Śląskiej. Organizacja i Zarządzanie*, 181. https://managementpapers.polsl.pl/wp-content/uploads/2023/12/181-Stecu%C5%82a.pdf

Stecuła, K. (2023b). Review of virtual reality applications applied in industrial practice. *Zeszyty Naukowe Politechniki Śląskiej. Organizacja i Zarządzanie*, 187. https://managementpapers.polsl.pl/wp-content/uploads/2024/01/187-Stecu%C5%82a.pdf

Stecuła, K. (2024). Analysis of asymmetric VR games—Steam platform case study. *Technology in Society, 78.*

Stecuła, K. (2025). Advancing education through virtual reality in the management and production engineering field of study. *Zeszyty Naukowe Politechniki Śląskiej. Organizacja i Zarządzanie*, 215. https://managementpapers.polsl.pl/wp-content/uploads/2025/05/215-Stecu%C5%82a.pdf

Stecuła, K., & Naramski, M. (2025). Virtual reality as a green tourism alternative: Social acceptance and perception. *Sustainability, 17*(17), 7722.

Stecuła, K., Wolniak, R., & Aydın, B. (2024). Technology development in online grocery shopping—From shopping services to virtual reality, metaverse, and smart devices: A review. *Foods, 13*(23), 3959.

Suess, C., & Maddock, J. (2025). Understanding the influence of window views, plantscapes, and green décor in virtual reality hospital rooms on simulated acute-care patients' stress recovery and relaxation responses. *Health Environments Research and Design Journal, 18*(3), 165–183.

Sultana, Z., & Kumar, D. (2024). Elevating 5G virtual reality projection screen platform with smart cities innovation and big data integration. *International Journal of Intelligent Systems and Applications in Engineering, 12*(3s), 141–150.

Tang, H. K. M., Lake, M. J., Foster, R. J., & Bezombes, F. A. (2025). Design, development and pilot of a realistic virtual reality application to analyse quick directional change in sport: Avatar cutting scenario with alterable parameters. *PLoS One, 20*(6), e0324941.

Thandayuthapani, S., & Thirumoorthi, P. (2025). Enhancing consumer engagement through augmented reality: AR's role in personalized shopping experiences, virtual product interaction, and sustainability. In: *Sustainable practices in the fashion and retail industry* (pp. 255–273). IGI Global.

Toh, E., Yoong, S. Q., Win Myint, O., Leong, Q. N., & Jiang, Y. (2025). Effect of virtual reality therapy on pain relief in sports rehabilitation for young adults: A systematic review and meta-analysis. *Annals of Physical and Rehabilitation Medicine, 68*(4), 101949.

Tsai, S.-H., Chao, C.-C., Lin, Y.-D., Wang, J.-H., & Liu, C. H. (2025). Immersive virtual reality cruise garden relaxation train: A novel BCI platform for depression. *IFMBE Proceedings, 118*, 15–31.

Vasudev, S., Kumari, P., Sawlani, J., & Sharma, P. (2025). Immersive urban narratives: Augmented reality data visualization for smart city engagement. *2025 IEEE International Students Conference on Electrical Electronics and Computer Science (SCEECS 2025).*

Vidal-Almela, S., Cadenas-Sanchez, C., & Ortega, F. B. (2025). Virtual reality sports to tackle pediatric obesity. *Nature Medicine, 31*(7), 2121–2122.

Vona, F., Schorlemmer, J., Stemann, J., Fischer, S., & Voigt-Antons, J.-N. (2025). Hands vs. controllers: Comparing user interactions in virtual reality shopping environments. *Proceedings of the 2025 IEEE Conference on Virtual Reality and 3D User Interfaces Abstracts and Workshops (VRW 2025)*, 1500–1501.

Wajid, M. S., Terashima-Marín, H., Zafar, A., Wajid, M. A., & Bhushan, B. (2025). *Digital twins for smart cities and urban planning: From virtual to reality.* CRC Press.

Wang, J., Qin, Y., Wu, Q., Sheng, B., & Li, H. (2025). An adaptive AI-based virtual reality sports system for adolescents with excess body weight: A randomized controlled trial. *Nature Medicine, 31*(7), 2255–2268.

Wieczorek, A., & Stecuła, K. (2025). Supporting sustainable demographic change: Augmented reality in men's sheds as a concept for active aging. *Sustainability, 17*(18), 8345.

Wiltenburg, R., Hurst, W., Mendoza, F. R., Krampe, C., & Tekinerdogan, B. (2025). Efficacy of smart city data layers in virtual reality for emergency evacuation behaviours. *Array, 27*, 100457.

Wolniak, R., Gajdzik, B., Grebski, M., Danel, R., & Grebski, W. (2024). Business models used in smart cities—Theoretical approach with examples of smart cities. *Smart Cities, 7*(4), 1626–1669.

Wolniak, R., & Stecuła, K. (2024). Artificial intelligence in smart cities—Applications, barriers, and future directions: A review. *Smart Cities, 7*(3), 1346–1389.

Wolniak, R., Stecuła, K., & Aydın, B. (2024). Digital transformation of grocery in-store shopping—Scanners, artificial intelligence, augmented reality and beyond: A review. *Foods, 13*(18), 2948.

Xing, H., Li, X., Liu, M., & Zhong, Y. (2025). Implementation of personalized customization and enhanced experiences for cultural tourism resources using genetic algorithm-based virtual reality technology. *Systems and Soft Computing, 7*, 200360.

Yue, W., Ma, K., Liu, Z., Wang, Y., & Shao, Y. (2025). Metaverse sports teaching: Virtual reality integration and innovation in college physical education courses. *Proceedings of the 2024 International Conference on Sports Technology and Performance Analysis (ICSTPA 2024)*, 489–493.

Zhang, L., Yu, H., & Xu, T. (2025). Simulation of interaction e-learning mode in political courses based on virtual reality and digital entertainment environment. *Entertainment Computing, 52*, 100857.

Zhang, Z. (2021). Research on the application of virtual reality technology in smart city construction. *ACM International Conference Proceeding Series*, 3470311.

CHAPTER 8

Limitations and Threats of Virtual and Augmented Reality

8.1 IDENTIFYING LIMITATIONS

Technologically, virtual reality (VR) and augmented reality (AR) are breaking through in the background of hardware vulnerability, software fragmentation, and limited interoperability. Hardware utilized to deliver immersed virtual experience depends upon the need for high computing, advanced graphics management, and rapid connectivity. Nevertheless, even premier-level systems are imperceptible without seams. Latency or reaction time of the system to the user is an intrinsic issue, which has a tendency to lead to loss of immersion in the majority of instances. Likewise, the rendering of complex scenes is strong and computationally intensive, excluding portability and performance. AR systems also include other challenges, i.e., digital superposition over dynamic real-world environments. Spatial detection and object tracking accuracy are normally sub-optimal and add misalignments degrading usability for safety-critical applications, i.e., surgical guidance or factory repair. Also, the absence of interoperable platforms has a splintered software platform that will not allow developers to create cross-compatible apps and will not allow mass adoption (Lee et al., 2025; Ramesh, 2025).

The social nature of VR and AR constrains provokes a paradox of duality: the technologies have the potential for introducing new ways of

DOI: 10.1201/9781003643395-8

connection, but may also hasten disconnection. Virtual worlds enable one to get attached in worlds that do not exist at all, and sometimes in lieu of actual socialization outside the world. Excessive use of this mode can be utilized to acclimatize human contact, especially in the young ones for whom virtual life could serve as a substitute for real human contact. Second, AR technologies, by inscribing digital information onto the textures of everyday life, are also likely to generate new sources of distraction and foster fragmented, disengaged attention. Public consumption of AR glasses, for example, is one to fret over in terms of sacrificing face-to-face contact but also to familiarize us with mediated over unmediated contact. In addition to this, the psychological effects of living in digitally created personas, frequently accompanied by personas used in real life, warp notions of self, identity construction, and social relationship continuity. Even though VR and AR have the ability to reduce distance, they envelop people near in physical proximity but far apart in the mind with walls of invisibility (Alfakir et al., 2025; Hu & Wei, 2025; Ransing et al., 2025).

The budgetary limitations of VR and AR are also paramount. Even as equipment is cheaper, costly equipment remains beyond the reach of vast segments of society. High-priced headsets, haptic feedback, and spatial sensors are resource-hungry, especially if their periodic updating will need to be carried out to maintain the pace of constantly evolving technology. Small businesses and schools can never afford access. Besides hardware, software application of VR and AR comes accompanied by expertise at the expense of designing, testing, and maintenance. Dislike for return on investment holds wide corporate uptake in check other than for entertainment and specialty industrial applications. Second, VR and AR business models are just starting out, with most companies experimenting with subscription models or content sites that will not survive in the long term. Its economic system is therefore precarious, and the level at which some massive companies dominate the market has sounded alarm bells regarding monopoly power and narrowed heterogeneity of innovation (Elsenbast et al., 2024; Gerwann et al., 2024; Stecuła, 2024; Stecuła, 2025).

Accessibility is similarly an inherent limitation, at least on the social equity aspects. While VR and AR are touted as global technologies that can level the playing field in education, health, or culture, it is not that straightforward. The rural-urban, developed-developing world, haves-have-nots digital divide persists. The astronomical expense, infrastructural requirements, and uneven availability of the net limit access to a relatively privileged group. Otherwise, not-accessing would also need to be considered not only financially, but physically and psychologically. Existing VR and AR are not yet accessibly equal to everyone. Blind, deaf, or motor-impaired individuals typically are faced with disabilities that keep them from achieving the same type of immersion or usability. Interfaces are still being designed very literally for the non-disabled, and inclusive design is still in its infancy but growing. If left unresolved, these inequalities are likely to go on perpetuating already widespread social inequalities by offering some accessible virtual and augmented realities and denying them to others (Díaz-Barrancas et al., 2024; Jonek-Kowalska & Wolniak, 2025; Sokołowska, 2024).

The medical limitations of VR and AR are also being ever better realized to be at the centre of recent controversy over their safe use. Prolonged use of VR headsets has resulted in motion sickness, eye strain, and headaches from sense-body mismatch. Reflected as cybersickness, the illness is experienced by a large number of users, restricting how long and how they can use them. Musculoskeletal pain, too, is caused by ergonomic design of equipment, mostly the neck and the back, because headsets are heavy and extremely uncomfortable to wear. Beyond physical distress, psychological effects of VR and AR use need to be stringently adhered to. Virtual worlds confuse real and simulated, and cause secondary dissociation or disordered spatial perception. In a few, particularly youth, sustained virtual world use can initiate addictive behaviour or exacerbate depression and anxiety disorders. AR use in everyday life, ambient overlaying of digital stimulation, also creates problems of cognitive overload, reduced attention spans, and long-term effects on memory and decision-making capacity. These are not only individual risks but also public health issues that must be regulated and controlled in an orderly fashion (Emilio et al., 2025; Gerwann et al., 2024; Mahmood et al., 2024; Sadeghian & Hassenzahl, 2021; Sharma et al., 2024; Skotnicka-Zasadzień & Wolniak, 2026; Rosak-Szyrocka and Wolniak, 2025; Jonek-Kowalska & Wolniak, 2023).

The absence of virtual and augmented reality uncovers the scene as greasy as it seems. The technical issues of latency, hardware cost, and software fragmentation converge with social issues of self, distraction, and isolation. Budget issues touch some but not all individuals and institutions and further reinforces digital divides over closing them. Access points highlight that inclusion is still an exclusive concern in AR and VR development. Lastly, the challenges in health issues show that there must be research in both the physical and psychological immersion aspects. Accepting these constraints does not downvalue the potential of VR and AR as revolutionary technologies but rather offers a platform on which development can act responsibly. Otherwise, use of these technologies can aggravate imbalances, disqualify human happiness, and solve more problems than problems. Conversely, acknowledgement and transcending these limitations can cause VR and AR to move in directions that are accessible, sustainable, and even beneficial to society (Choi et al., 2023; Salkovska et al., 2024; Stecuła & Naramski, 2025; Xu et al., 2023; Stecuła, 2024).

Table 8.1 below provides virtual and augmented reality's most serious limitations' definition for five categories: technological, social, economic, accessibility, and health. Each category captures concerns that currently hinder mainstream adoption, ranging from hardware inefficiencies and high costs to issues of marginalization, exclusion, social isolation, and health-related risks. Placing the concerns within these categories, the table facilitates an understanding of multi-disciplinary challenges which need to be overcome prior to VR and AR becoming possible and realizable aspects of society.

TABLE 8.1 Limitations of Virtual and Augmented Reality

Limitation	Description
Technological	Both AR and VR require low-latency networking, high-performance computing, and good graphics, but the maturity of the current hardware is to be unable to enable imperceptible immersion. The battery runs short, the hardware is bulky, and AR space tracking is imprecise. Software fragmentation and lack of interoperability keep it back from adoption.
Social	These technologies isolate as they replace mediating experience with human interaction. AR distracts the public and reduces face-to-face interaction. Virtual construction of identity also annihilates authenticity and psychological well-being.
Economic	Premium headsets, haptic peripherals, and sensors are barriers to entry for individual consumers, small and medium businesses, and schools. Software development takes a long time and is a difficult process. Uncertainty of return on investment and volatility in markets are barriers to entry for broader adoption from other niches and entertainment markets.
Accessibility	VR and AR are currently out of reach to the masses with expensive devices, infrastructural requirements, and unequal internet access. Disabled populations mainly enjoy access to contemporary hardware since it is designed for non-disabled persons. This strengthens digital divides in and between places and social groups.
Health	Long-term use of VR headsets has been shown to lead to motion sickness, eye strain, and neck strain as a result of sensory incongruences and ergonomics. VR and AR also carry the potential risk of mental overload, attention fragmentation, and addiction. Psychological consequences in terms of dissociation or fear are also a concern.

Source: Author's own work.

8.2 IDENTIFYING THREATS

Table 8.2 lists the following worry for danger from VR and AR, ranging from data privacy and isolation to administrative ambiguity and ecological sustainability. A quick glance at all is that dangers are not on the sidelines but at the centre of how VR and AR are created and used responsibly.

TABLE 8.2 Threats Connected with Virtual and Augmented Reality

Threat	Description
Data privacy and security	The AR and VR technologies record body motion, biometric data, and environment, which can be utilized deceptively to monitor, profile, or act directly. Cyber attacks and unlawful intrusion into people's information remain an issue.
Social Isolation	Excessive use of virtual worlds may reduce physical-world engagement, undermine social coherence, and lead to estrangement or alienation, especially among youth.
Economic Dependence	They are hostage to expensive proprietary platforms, and they have more opportunity to be victims of monopoly exploitation, lock-in subscription, and unlawful suspension of service.
Health and Well-Being	Prolonged use of headsets causes motion sickness, eye fatigue, musculoskeletal discomfort, and mental fatigue. Psychologically, they are susceptible to addiction, dissociation, and nervousness due to the breakdown of boundaries between virtual and real.
Ethical and Societal Risks	Immersive simulations are even more vulnerable to the spread of disinformation, propaganda, or hate speech than mass media. There are also virtual harassment, exploitation, or stereotyping issues of a moral nature.
Digital Divide and Exclusion	Exclusionary design, cost, and infrastructure needs exclude disadvantage groups, enhancing world inequalities and ensuring non-equal access to digital opportunities.
Cognitive and Behavioural Manipulation	VR and AR experience can be engineered to quietly influence decision-making, action, or buying choice and invade autonomy and informed consent.
Legal and Regulatory Challenges	Increased innovation in immersive technology is occurring faster than the regulatory system, raising concerns over intellectual property rights, responsibility for damages, and regulation of the virtual world.
Workplace and Productivity Risks	Application of VR/AR in working environments may result in increased surveillance and monitoring of workers, inappropriate application of measurement of work output, and further exacerbation of uncertainties over privacy and autonomy of work.
Environmental Impact	Manufacturing of VR/AR equipment entails power-driven processes and creation of e-wastes, and sustainability issues in case of short or poorly recyclable product lives.

Source: Author's own work.

The gravest danger is to data security and privacy. AR and VR headsets are stockpiling orders of magnitude previously unimagined for consumer hardware. They're monitoring explicit movement such as hand and head rotation, and subconscious biometric movement, where the eyelids blink and even mental states inferred from patterns of use. Glasses and headsets, in the AR world, continuously scan their environment as a quest to present virtual information, virtually transforming individuals' offices, houses, and public areas into high-density three-dimensional maps. Such sensitivity and richness equate to huge vulnerabilities. The exploitation can be carried out by business exploitation for systemic advertising, illegal government surveillance, or criminal exploitation. The threat lies not just in potential intrusions of information but also in commodification of spaces where people are subject to continuous monitoring and surveillance without explicit consent. This goes against conventional readings of privacy, autonomy, and ownership of information in public as opposed to private spaces (Baldoni et al., 2025; Marino et al., 2025; Szymanek et al., 2025).

Social exclusion is also a huge risk, even though VR and AR are being touted as unifying factors. The paradox is that as the immersive nature enables interactive distance, it simultaneously reduces face-to-face contact. Virtual communication as a substitute for direct human contact alters communication and social way of life in the long term. Long-term replacement will be accompanied by reduced empathy, loneliness, and isolation between youth and young adults. Moreover, the embedding of AR in daily life can also violate attention and continuity of social interaction. Different incoming signals, overlays, or advertisements shown through AR glasses, for instance, can destabilize the quality of face-to-face interaction and reduce presence in bodily, physical-world social environments. These trends suggest that, unbridled, immersive tech is driving a shift from superficial to human connection (Klotzsche et al., 2024; O'Neill et al., 2024; Szymańska, 2025).

Financial hazards of VR and AR echo with system dependence on monolithic corporations. Immersive tech ecosystems are goliaths of a few multinational corporations that dominate hardware manufacturing, software ecosystems, and content platforms to sell. This focus creates monopoly power, in that the consumers are held captive within proprietary platforms, which expose them to subscription lock-in and reduced cross-platform portability. To the firms, it translates into insecurity and exposure to dependence upon capricious corporate whims. When the market-leading vendor introduces dropped services or introduces a change in its framework, the firms who have invested heavily in the technology face being financially and operationally hurt severely. It cannot be any other way, it's too much for small businesses because VR and AR technology is data- and capital-hungry. Economically, concentration not just heightens exposure but chokes diversity and development in the ecosystem Ramaseri-Chandra & Pothana, 2024; Wieczorek & Stecuła, 2025).

Health and well-being risks are a multi-dimensional threat. Physically, extended use of VR headsets causes eyestrain, motion sickness, musculoskeletal pain, and fatigue. They result from contradictions in body perception, weight, ergonomics, and sensory input of current devices. Cognitive aspects include the fact that both AR and VR subject users to information overload since the brain has to process so much audio and visual information simultaneously. Psychologically, presence in the computer world brings about the consequence of blurring fiction and reality and thus dissociation or disorientation. Excess also fosters reliance, replacing offline work, social bonding, or well-being with online consumption. Another increasing worry is that such adverse effects actually exacerbate anxiety, depression, or other sicknesses, particularly among vulnerable groups. The threat that immersive technology poses to not only physical but also psychic well-being positions health in a front-line zone of vulnerability that needs to be tightly defended (Lu et al., 2023; Prabakeran et al., 2024).

Damage of social and ethical types can be inflicted. Virtual and augmented reality environments can be used to manipulate with the aim of controlling, to deceive, or to exploit. Experiential disinformation or interactive propaganda will likely be more persuasive than media since it is experiential. Interactive campaigns can invite users into to-be-experienced simulations with continued delivered narratives and close-down critical distance, thereby leaving the user open to exploitation. Lastly, anonymity and virtual world immersion provide fertile grounds for exploitation, abuse, or harassment. Questions of accountability come into play in these cases: when dangerous behaviour is being engaged in within a completely virtual space, it is unclear who and what to hold accountable and why. Questions of ethics also come into play within the content creation space, where content creators are inadvertently recreating stereotypes, biases, or discriminatory behaviours, possibly indicating inequalities in online culture. In spite of strict ethical norms, VR and AR rather than curing social ills can perpetuate them (Abuhashish & Sunar, 2025; Lee & Seals, 2025; Stecuła, 2023a; Stecuła, 2023b).

Digital divide and exclusion also mirror structural imbalances between immersive technologies. Availability of VR and AR even where they are possible is underfunded by expensive hardware, broadband, and elitist architecture. Rural or underdeveloped communities are consciously excluded by infrastructural shortages. Likewise, the disabled will be stuck with record-setting disenfranchisement because most interfaces make the assumption of able-bodied users. Exclusionary tendencies like these will keep opening chasms by providing advantage groups with access to high-tech digital possibilities while leaving others behind. Instead of democratizing power, VR and AR will aim to further increase gaps among engaged and disengaged individuals and thereby deepen socio-economic divisions (Borhani & Ortega, 2023; Ramesh, 2025).

The second threat is behavioural and cognitive manipulation. The presence-based nature of VR and AR gives developers the luxury of designing experiences to influence consumers' decisions and moods involuntarily and automatically as they do on emotion. Gameworlds instantiate gameplay norms through reward schemata, and AR advertising instantiates commodities into ordinary people's line of sight. Both are manipulations that circumvent ordinary critical defense strategies, and issues of autonomy and informed consent arise. Such a capability for this degree of intimate behaviour control is an incredible tool but an incredible danger if mobilized for political or commercial purpose.

Regulatory and judicial uncertainty are the genesis of most of the foregoing risk. Technology change is much faster than change in general principles of common law. Intellectual property rights in virtual worlds, tort liability for injury to virtual worlds, or property rights in a virtual economy are all unanswered questions. Uncertainty is not just bad news for users, who enjoy fewer avenues to appeal if injured, but for companies, too, who are working without the solid regulative blanket. The absence of strict legal rules allows evil behaviour without supervision and erodes faith in the system overall (Han & Clevenger, 2025).

Employment of VR and AR in the workplace creates workplace and productivity risks. Whereas immersion technology promises to power more-scale training and co-design, it also promises new forms of manipulation and surveillance. Such technology can be employed by managers to monitor workers' behaviour, rate of productivity, and even physiological state in real time. Such practices can empty working relationships of privacy, autonomy, and trust. Apart from this, the rapid implementation of VR and AR at the workplace is also accompanied by the impact of disenfranchising susceptible workers who lack the necessary technological expertise and the associated hardware, thus leading to disparities in labour markets (Kaushik et al., 2024; Quintana et al., 2023; Wang et al., 2025). The danger lies in long-term environmental sustainability.

Production of VR and AR technology is highly energy-intensive with rare earth elements, production involving high-energy processes, and global supply chains. The fast cycle of innovation is reducing the life of devices, generating colossal e-waste. In the absence of recycling machines and green manufacturing processes as a priority, environmental destruction through the application of VR and AR will render all sustainability work on this earth useless. This problem reflects the broader power of immersive technologies to reshape not only social and virtual spaces, but also ecological systems, by influencing how the world is organized and governed (Baldoni et al., 2025; Szymanek et al., 2025).

8.3 ADDRESSING CHALLENGES AND PROPOSING SOLUTIONS

Scientists investigating VR and AR came to the conclusion that there are some common issues to be solved to a great extent. Solutions of technological, social, regulatory, and ethical nature have been offered first of all, which have some way of performing VR and AR responsibly. In this essay, the most important issues of VR and AR and some combinations of solutions trying to minimize their risk and maximize their utility are touched upon.

Technology immaturity is the biggest issue immersive technology currently confronts. VR and AR depend on high-computation requirements, high-fidelity graphics, accurate motion tracking, and buffer-less networking. Current hardware always falls short of those requirements. User perception of latency is when timing delay between input and output destroys immersion. So also do AR apps often struggle with environmental mapping and spatial awareness, creating misapplied overlays that destroy usability in business application arenas like surgery or industrial service work. It comes with hardware design limitation: headsets remain very heavy, ergonomically second-best, and hamstrung with anemic battery life. Solutions along this path are more engineering- and tech-centric. Also on the agenda is a move towards lighter hardware at the lead fueled by 5G and wirelessness, and superior tracking algorithms based on artificial intelligence. Cloud rendering investment is also on the agenda, decoupling computation from mobiles to battered networks and hence cutting latency while opening up the capabilities of mobile applications. Interoperability cutting fragmentation and providing an integrated ecosystem by hardware and software vendors also enter the picture here (Garg et al., 2025; Samala et al., 2025).

Social isolation and replacement behaviour is yet another very relevant serious issue. While immersive technologies make distant interaction real, they also have the ability to decrease face-to-face interaction and other social engagement. A person, especially the youth, can substitute mediated relations with real life relations, which leads to additional isolation and alienation. AR application to daily living is also sure to be continuously deflected and divided attention. Intervention will thus have to transcend the territory of technology to social education and ethical space in this respect. Programmers can design systems in which collaboration is a primary objective rather than an occasional, mission-specific feature, enabling seamless teamwork in virtual and augmented environments. Educational software also brings parents, teachers, and students to awareness about balance between immersed activity and off-line engagement. Ethical principles like 'digital well-being by design' can be incorporated into VR/AR platforms at the design stage in order to prevent overuse and addiction. Social researchers emphasize the need for hybrid, complementary experiences rather than alternative forms of interaction that displace face-to-face contact; technology should complement communal relations, not replace them (Kim et al., 2025; Musarat et al., 2025; Smith et al., 2025; Wolniak & Stecuła, 2024; Wolniak et al., 2024a; Wolniak et al., 2024b).

Adoption is also challenging because of economic constraints. Costly limiting hardware, specialized software, and skill set limitation restrict usage to most individuals, companies, and organizations. Small and medium businesses, schools, and healthcare facilities cannot afford to adopt immersive technologies in spite of the hype. In addition, concentration of market power in the form of a few giant multinationals creates dependence and monopolistic risks. Solutions involve establishing open-source communities that minimize dependence on closed frames and lower the entry barrier. Public funding schemes, especially for education and healthcare, can render immersive technology affordable. Public-private collaborations, grants, and subsidies are also being experimented by some city and national governments as a means of implementing VR and AR into school networks or hospital systems. Cost-affordable business models that are within reach should also be accessible: rather than high-cost ownership, subscription or shared-access plans can provide access for small organizations to immersive technology. Finally, the democratization of VR and AR depends on achieving a balance between innovation on the one hand and cost reduction and scalability on the other (Felszeghy et al., 2025; Geris & Kulaksiz, 2025; Schaumburg et al., 2025).

Accessibility is another related concern, all too easily pushed under the mat when populist tub-thumping occurs. The overwhelming majority of VR and AR platforms are accessible to the non-disabled but not to visually impaired, hearing impaired, or mobility impaired. Additionally, international variation in access to the internet, the cost of hardware, and digital competence continues to perpetuate the digital divide. Something is needed so that the immersive technologies are not additive to inequalities but possess a counter-effect action. This is achieved by including considerations for accessibility from the beginning in the design process. Compliance with universal design principles will help make it accessible with assistive technology, voice control, haptic feedback, and adaptive interfaces available with evolving user needs. Initiatives to roll out broadband infrastructure, especially to low-income or rural communities, are equally important. Digital skills modules will also make marginalized communities stakeholders of immersive environments. Beyond that, business and regulatory demands to develop accessibility standards can drive organizations towards inclusivity as an unwelcome truth of innovation (Nakić et al., 2025; Pandrangi et al., 2025).

Health and well-being issues are being placed at the forefront as underlying drivers to continued take-up. Physiologically, VR headsets are also discovered to induce motion sickness, visual fatigue, and musculoskeletal discomfort as a result of prolonged use. Cognitively, virtual experience can easily be associated with overwhelm attention or interfere memory. Psychologically, there will be addiction, dissociation, or increased anxiety from prolonged use of virtual worlds. There are two ways to these health risks. From the engineering point of view, increased refresh rates, decreased latency, and ergonomics will avoid physical fatigue. Psychologically, open policies of responsible usage, e.g., time limits on screen usage, will manage exposure. Warning mechanisms that detect over-use and suggest a break can also be added by developers. Multi-disciplinary studies by neuroscientists, psychologists, and technologists can also enlighten us regarding the impact of immersion on the human body and brain so that product policy and design could be altered as a result. Health-conscious design will act as the custodian for defending the user because it maximizes VR and AR adoption (Palakurti, 2025; Zhang & Yao, 2025).

Social concerns and ethics is the second. Immersive technology can mislead, deceive, or spread objectionable content. Since VR and AR are more realistic than other media types, their impacts on cognition and behaviour can be powerful. Anonymity in cyberspace can be leveraged to facilitate harassment or exploitation. Solutions, therefore, must address matters of governance and accountability. Ethical contribution must inform content creation for immersive content so that the applications do not play a role in perpetuating stereotypes, reinforcing negative ideologies, or taking advantage of vulnerable consumers. Immersive worlds should be governed so that the developer, content provider, and platform provider roles are established. International harmony will be inevitable due to the international character of immersive technologies which cannot be brought under control by rival national law. Another fascinating question is that of creating digital literacy training allowing the users to critically identify immersive experience and react to manipulative use (Chamusca et al., 2025).

Regulatory and legislative issues cut across some of the above issues. Immersive environments generate new intellectual property, liability, and ownership issues. Virtual property created or bought in VR worlds or injury caused by harassment or invasion in worlds in immersive worlds is prone to litigation. Institutional systems prevalent today fall short in addressing this extent of complexity. Countermeasures need evolving government and technology. Economic and social law has to acknowledge virtual worlds as real space for economic and social existence in order to offer protection and duties there. User needs for transparency, accountability, and security should be addressed through multi-stakeholder consultation involving governments, industry, and civil society. International organizations such as the United Nations and the European Union can further support the development of harmonized frameworks and common standards, thereby reducing regulatory fragmentation (Jiang et al., 2025; Selvam & Gnana Piragasam, 2025).

Sustainability of the environment is not a short-term but ever-increasing issue at the core of the issue of immersive technologies. VR and AR hardware manufacturing is dependent on electricity and relies on the use of rare earth elements (Table 8.3). Repeated revising and short production cycles generate e-waste and cause environmental degradation. The measures involve turning consumption and production processes around. Circular economy applications such as renovation programs and

TABLE 8.3 Addressing Challenges of VR and AR and Proposing Solutions

Challenge	Description of Challenge	Proposed Solutions
Technological	Low battery life, heavy headsets, compromised spatial tracking, and malfunctioning software environments reduce use and immersion.	Lightweight and ergonomic device design; cloud rendering to minimize local processing load; use of AI to improve tracking; use of 5G/6G for low-latency networking; creation of industry-wide interoperability standards.
Social	Risk of social isolation, divided attention, and substitution of embodied with mediated experience.	Encouragement of social VR/AR applications; incorporation of 'digital well-being by design' principles; hybrid experiences with physical and digital interaction; consumer education about balanced use.
Economic	High equipment cost, custom software, and support discourages adoption in schools, SMEs, and healthcare. Market concentration promotes monopolization.	Open-source platforms; shared-access/subscription business models; government grants and subsidies; public-private partnerships; enabling small developers with innovation funding.
Accessibility	Exclusion of disabled people and groups where provision is lacking or where broadband is not available. Interfaces continuing to be constructed for majority able-bodied users.	Universal design practices implementation; voice, haptic feedback assistive technology support; development of mass-market entry devices; broadband infrastructure upgrade; digital literacy initiatives.
Health and Well-Being	Motion sickness, eyestrain, musculoskeletal issues, mental overload, and psychological risks (addiction, dissociation, anxiety).	Enhanced ergonomics and refresh rates; patient education guides for safe usage; internal notification of overuse; multi-disciplinary research on long-term health consequences; wellness features integrated into platforms.
Ethical and Societal	Abuse risk for manipulation, mis-information, harassment, and dissemination of stereotypes.	Creation of ethical standards for immersive content; regulation of misleading or dangerous uses; standards for content moderation policies; promotion of international cooperation towards the creation of ethical standards.

(Continued)

TABLE 8.3 Addressing Challenges of VR and AR and Proposing Solutions (*Continued*)

Challenge	Description of Challenge	Proposed Solutions
Legal and Regulatory	Intellectual property, liability, virtual asset ownership, and protection of users in immersive environments are still to be settled.	Adaptive legal frameworks; harmonized global norms; clear definitions of liability; regulatory authorities for immersive technologies; clear rights and responsibilities of users.
Workplace	Surveillance, over-monitoring, and differential uptake across various categories of workers are dangerous.	Clarity in privacy legislation in action; clarity in data use; policy of voluntary uptake; staff training; fair access to immersive workplace technologies.
Environmental	Waste energy in the manufacture of headsets, constant upgrading, and finite lifespan of devices all result in e-waste.	Application of circular economy principles (device refurbishment, reusing, recycling); green materials and device modularity for sustainability; policy incentives for green manufacturing; consumer awareness campaigns.

Source: Author's own work.

recycling can prevent wastage. Low energy consumption, modularity, and longevity environment design principles can be adhered to by designers. Policymakers can encourage the integration of sustainability attributes through subsidies, certification schemes, and legislation. Consumer literacy is also significant: purchase, replacement, and recycling of immersion equipment offsets the ecological footprint of immersive technology (Bhambri, 2025; Oztank & Balcisoy, 2025; Qian et al., 2025). The sustainability agenda should not be put on hold if VR and AR technologies become a dominant industry priority; otherwise, the sector may advance at the expense of broader societal and environmental well-being.

BIBLIOGRAPHY

Abuhashish, F., & Sunar, M. S. (2025). Augmented reality in cybersecurity: Enhancing threat intelligence and protecting critical infrastructure. In: *Complexities and challenges for securing digital assets and infrastructure* (pp. 291–321). IGI Global.

Alfakir, R., Holt, G., Mason, L., Weisbrod, L., & Trehub, A. (2025). Promoting empathy in audiology education through virtual reality and tactile technologies: A pilot study for patient-centered care in individuals with hearing loss and manual dexterity limitations. *Medical Science Educator*. https://link.springer.com/article/10.1007/s40670-025-02403-x

Azrour, M., Hossain, I., & Mahmudul Haque, A. K. M. (2025). Addressing environmental challenges with AI, robotics, and augmented reality. In: *Addressing environmental challenges with AI, robotics and augmented reality* (pp. 1–424). Springer Nature.

Baldoni, S., Benhamadi, S., Chiariotti, F., Zorzi, M., & Battisti, F. (2025). Movement- and traffic-based user identification in commercial virtual reality applications: Threats and opportunities. *Proceedings of the 2025 IEEE Conference on Virtual Reality and 3D User Interfaces (VR 2025)*, 72–81.

Bhambri, P. (2025). Virtual and augmented reality for immersive learning experiences: Practical applications and challenges. In: *Architecting the digital future: Platforms design and application* (pp. 17–46). IGI Global.

Borhani, Z., & Ortega, F. R. (2023). A virtual reality system for gender swapping to increase empathy against stereotype threats in computer science job interviews. *Proceedings of the 2023 IEEE International Symposium on Mixed and Augmented Reality Adjunct (ISMAR Adjunct 2023)*, 648–649.

Chamusca, I. L., Pagano, T. P., Loureiro, R. B., Murari, T. B., & Winkler, I. (2025). Enhancing context awareness in augmented reality environments: A review of opportunities and challenges in object and action recognition through machine learning. *Discover Applied Sciences*, *7*(7), 750.

Choi, D. H., Kim, S.-K., & Kim, S. (2023). Real-time hole-filling in mobile augmented reality gaming: A novel algorithm to overcome depth sensor limitations. *Traitement du Signal*, *40*(4), 1377–1384.

Diao, X., Rownak, M. R., & Smidts, C. (2024). Design and implementation of a virtual reality environment for human reliability assessment under physical security threats. *Proceedings of the Pacific Basin Nuclear Conference (PBNC 2024)*, 716–725.

Díaz-Barrancas, F., Flores-Martin, D., & Berrocal, J. (2024). Training a neural network on virtual reality devices: Challenges and limitations. *Proceedings of the 2024 IEEE Conference on Virtual Reality and 3D User Interfaces Abstracts and Workshops (VRW 2024)*, 787–788.

Elsenbast, C., Dahlmann, P., & Schnier, D. (2024). Virtual team training with mixed reality and virtual reality—Benefits and limitations illustrated on the example of two paramedic classes. *Multimedia Tools and Applications*, *83*(23), 63871–63895.

Emilio, P., Thiha, R. G., Jegors, T., & Chen, Z. (2025). Bridging the sensory gap in augmented reality: A systematic review on the impact of sensory limitations on consumer purchase intent. *IET Conference Proceedings*, *2025*(15), 65–67.

Felszeghy, S., Mutluay, M., Liukkonen, M., Bencharit, S., & Quinn, B. (2025). Benefits and challenges of the integration of haptics-enhanced virtual reality training within dental curricula. *Journal of Dental Education*, *89*(7), 1070–1083.

Garg, V., Vinod, G., Singh, S., & Kant, V. (2025). Human reliability analysis for internal fire event with virtual reality: Conceptualization and challenges. *Nuclear Engineering and Design*, *444*, 114379.

Geris, A., & Kulaksiz, T. (2025). Navigating the integration of virtual reality in education: NLP-based perspectives, benefits, and challenges. In: *Immersive learning in teacher education simulated environments: Tools and practices* (pp. 59–100). IGI Global.

Gerwann, S., Baetzner, A. S., & Hill, Y. (2024). Immersive virtual reality and augmented virtuality in sport and performance psychology: Opportunities, current limitations, and practical recommendations. *Sport, Exercise and Performance Psychology, 14*(1), 268–283.

Han, B., & Clevenger, C. M. (2025). Augmented reality and mixed reality implementation status survey in the architecture, engineering, and construction industry: Benefits, limitations, and future directions. *Journal of Architectural Engineering, 31*(2), 04025017.

Hu, X., & Wei, B. (2025). Virtual reality in obesity management: Addressing key limitations in clinical application. *Journal of Medical Internet Research, 27*, e68269.

Jiang, H., Zhu, D., Chugh, R., Turnbull, D., & Jin, W. (2025). Virtual reality and augmented reality-supported k-12 STEM learning: Trends, advantages and challenges. *Education and Information Technologies, 30*(9), 12827–12863.

Jonek-Kowalska, I., & Wolniak, R. (2023). *Smart cities in Poland: Towards sustainability and a better quality of life?* Routledge.

Jonek-Kowalska, I., & Wolniak, R. (2025). *Exclusion in smart cities: Assessment and strategy for strengthening resilience.* Routledge.

Kaushik, P., Vignesh Kumar, P., Dowlath Nisha, S., Bano, S., & Gupta, S. (2024). Exploring the efficacy of virtual reality technology in enhancing emotional regulation and therapy: Benefits, limitations, and future directions. *Proceedings of the IEEE 1st International Conference on Advances in Computing Communication and Networking (ICAC2N 2024)*, 1444–1449.

Kim, H., Han, E. C., Muntz, P. B., & Kemp, J. (2025). Virtual reality in the treatment of anxiety-related disorders: A review of the innovations, challenges, and clinical implications. *Current Psychiatry Reports, 27*(9), 519–528.

Klotzsche, F., Motyka, P., Molak, A., Fajnerová, I., & Gaebler, M. (2024). No cardiac phase bias for threat-related distance perception under naturalistic conditions in immersive virtual reality. *Royal Society Open Science, 11*(10), 241072.

Lee, C., Redden, L., Eng, V., & Eadie, B. (2025). Luminance and thresholding limitations of virtual reality headsets for visual field testing. *PLoS One, 20*(9), e0332795.

Lee, J., & Seals, A. (2025). Capturing invisible health threats through augmented reality-based public health messaging: Focusing on perceived realism, issue involvement, and message attitudes. *Behaviour and Information Technology, 44*(11), 2620–2636.

Lu, J., Kemmerer, S. K., Riecke, L., & De Gelder, B. (2023). Early threat perception is independent of later cognitive and behavioral control: A virtual reality–EEG–ECG study. *Cerebral Cortex, 33*(13), 8748–8758.

Mahmood, A., Butler, B., Zen, H., & Jennings, B. (2024). Key early verticals: Challenges and limitations in implementation of augmented reality. In: *Encyclopedia of computer graphics and games* (pp. 1025–1029). https://link.springer.com/rwe/10.1007/978-3-319-08234-9_273-1

Marino, C. E., Rjabtsenkov, P., Sharp, C., Zhu, X., & Suarez-Jimenez, B. (2025). Using virtual reality to study spatial mapping and threat learning. *NPJ Science of Learning, 10*(1), 14.

Musarat, M. A., Khan, A. M., Alaloul, W. S., Ayub, S., & Bérenger, M. (2025). Inspecting the benefits and challenges of virtual reality in construction projects of Malaysian industry. *Innovative Infrastructure Solutions, 10*(8), 350.

Nakić, J., Rogulj, I., & Maleš, L. (2025). Rethinking the Bebras challenge in virtual reality: Implementation and usability study of a computational thinking game. *Multimodal Technologies and Interaction, 9*(6), 60.

O'Neill, D., Titmus, M., Lamont, W., Perimal, E., & Di Pietro, F. (2024). Validating the threat of a virtual reality clinical environment: A mixed methods study. *Applied Sciences, 14*(21), 10009.

Oztank, F., & Balcisoy, S. (2025). Towards extended reality in emergency response: Guidelines and challenges for first responder friendly augmented interfaces. *Computer Animation and Virtual Worlds, 36*(3), e70056.

Palakurti, A. (2025). Harnessing AI and augmented reality for academic research for innovations, applications, and ethical challenges. In: *Revolutionizing academic research with AI and augmented reality* (pp. 33–66). IGI Global.

Pandrangi, V. C., Araujo, A. V., Buncke, M., Geltzeiler, M., & Li, R. J. (2025). Postoperative implementation of virtual reality and wearable devices: Opportunities and challenges. *Laryngoscope, 135*(6), 2044–2050.

Prabakeran, S., Sethukarasi, T., & Indumathi, V. (2024). Threat matrix: Stepping into the future of cybersecurity training with augmented reality. In: *Navigating the augmented and virtual frontiers in engineering* (pp. 239–261). IGI Global. https://www.igi-global.com/chapter/threat-matrix/351543

Qian, Y., Yang, Z., Chen, S.-C., Lin, C.-L., & Wu, S.-T. (2025). Power consumption of light engines for emerging augmented reality glasses: Perspectives and challenges. *Advanced Photonics, 7*(3), 034001.

Quintana, D., Rodríguez, A., & Boada, I. (2023). Limitations and solutions of low cost virtual reality mirror therapy for post-stroke patients. *Scientific Reports, 13*(1), 14780.

Ramaseri-Chandra, A. N., & Pothana, P. (2024). Cybersecurity threats in virtual reality environments: A literature review. In: *Proceedings of the cyber awareness and research symposium (CARS 2024)*. https://ieeexplore.ieee.org/document/10778838

Ramesh, M. R. (2025). Navigating the virtual road: Challenges and limitations of virtual and augmented reality in the automobile industry. In: *Virtual and augmented reality applications in the automobile industry* (pp. 361–376). IGI Global.

Ransing, V., Park, J., Ye, Y., Du, J., & Srinivasan, D. (2025). Efficacy of virtual reality-based approach for users to understand the potential benefits and limitations of using exoskeletons. *Human Factors*. https://journals.sagepub.com/doi/10.1177/00187208251346627

Rosak-Szyrocka, J., & Wolniak, R. (2025). *Smart universities in smart cities: Shaping the future of education and urban innovation*. Routledge.

Sadeghian, S., & Hassenzahl, M. (2021). From limitations to superpowers: A design approach to better focus on the possibilities of virtual reality to augment human capabilities. *Proceedings of the 2021 ACM Designing Interactive Systems Conference (DIS 2021)*, 180–189.

Salkovska, J., Batraga, A., Kaibe, L., Minskere, L., & Kellerte, K. (2024). The use of augmented reality tools in marketing: Opportunities and limitations in the Latvian market. In: *Lecture notes in networks and systems* (pp. 43–53). Springer.

Samala, A. D., Rawas, S., Rahmadika, S., Fikri, R., & Sandra, R. P. (2025). Virtual reality in education: Global trends, challenges, and impacts—Game changer or passing trend? *Discover Education*, *4*(1), 229.

Schaumburg, M., Imtiaz, A., Zhou, R., Wolbers, T., & Segen, V. (2025). Immersive virtual reality for older adults: Challenges and solutions in basic research and clinical applications. *Ageing Research Reviews*, *109*, 102771.

Selvam, B., & Gnana Piragasam, G. A. (2025). Augmented reality's potential for addressing writing challenges in students with learning disabilities. *British Journal of Special Education*, *52*(2), 147–156.

Sharma, N., Mallela, A. N., Khan, T., Biehl, J., & Andrews, E. G. (2024). Evolution of the meta-neurosurgeon: A systematic review of the current technical capabilities, limitations, and applications of augmented reality in neurosurgery. *Surgical Neurology International*, *15*. https://pubmed.ncbi.nlm.nih.gov/38742013/

Skotnicka-Zasadzień, B., & Wolniak, R. (2026). Artificial intelligence in public administration. In: *Routledge studies in innovation, organizations and technology*. Routledge. https://www.routledge.com/Artificial-Intelligence-in-Public-Administration/Skotnicka-Zasadzien-Wolniak/p/book/9781032879284

Smith, L. C., Vernal, D. L., Mariegaard, L. S., Nordentoft, M., & Glenthøj, L. B. (2025). Immersive virtual reality-assisted therapy targeting persistent auditory verbal hallucinations in patients diagnosed with schizophrenia spectrum disorders in Denmark: The challenge assessor-masked, randomised clinical trial. *Lancet Psychiatry*, *12*(8), 557–567.

Sokołowska, B. (2024). Being in virtual reality and its influence on brain health—An overview of benefits, limitations and prospects. *Brain Sciences*, *14*(1), 72.

Stecuła, K. (2023a). Identifying diverse uses of virtual reality in higher education and exploring perceptions of VR in the chosen field. *Zeszyty Naukowe Politechniki Śląskiej. Organizacja i Zarządzanie*. https://managementpapers.polsl.pl/wp-content/uploads/2023/12/181-Stecu%C5%82a.pdf

Stecuła, K. (2023b). Review of virtual reality applications applied in industrial practice. *Zeszyty Naukowe Politechniki Śląskiej. Organizacja i Zarządzanie.* https://managementpapers.polsl.pl/wp-content/uploads/2024/01/187-Stecu%C5%82a.pdf

Stecuła, K. (2024). Analysis of asymmetric VR games—Steam platform case study. *Technology in Society, 78*, 102673.

Stecuła, K. (2025). Advancing education through virtual reality in the management and production engineering field of study. *Zeszyty Naukowe Politechniki Śląskiej. Organizacja i Zarządzanie.* https://managementpapers.polsl.pl/wp-content/uploads/2025/05/215-Stecu%C5%82a.pdf

Stecuła, K., & Naramski, M. (2025). Virtual reality as a green tourism alternative: Social acceptance and perception. *Sustainability, 17*(17), 7722.

Stecuła, K., Wolniak, R., & Aydın, B. (2024). Technology development in online grocery shopping—From shopping services to virtual reality, metaverse, and smart devices: A review. *Foods, 13*(23), 3959.

Szymanek, P., Nenadalová, J., & Van Leeuwen, N. (2025). Revisiting feeling of threat and agency detection: A preregistered virtual reality study. *Psychology of Religion and Spirituality, 17*(3), 161–174.

Szymańska, A. (2025). The Proteus effect in virtual reality: Examining stereotype threat through gender cues. *Proceedings of the 2025 IEEE Conference on Virtual Reality and 3D User Interfaces Abstracts and Workshops (VRW 2025)*, 1580–1581.

Wang, V. L., Cheng, L. G., Zhu, T. S., Patnaik, S., & Tarin, T. (2025). Current applications and limitations of augmented reality in urological surgery: A practical primer and 'state of the field. *Current Urology Reports, 26*(1), 56.

Wieczorek, A., & Stecuła, K. (2025). Supporting sustainable demographic change: Augmented reality in men's sheds as a concept for active aging. *Sustainability, 17*(18), 8345.

Wolniak, R., Gajdzik, B., Grebski, M., Danel, R., & Grebski, W. (2024a). Business models used in smart cities—Theoretical approach with examples of smart cities. *Smart Cities, 7*(4), 1626–1669.

Wolniak, R., & Stecuła, K. (2024). Artificial intelligence in smart cities—Applications, barriers, and future directions: A review. *Smart Cities, 7*(3), 1346–1389.

Wolniak, R., Stecuła, K., & Aydın, B. (2024b). Digital transformation of grocery in-store shopping—Scanners, artificial intelligence, augmented reality and beyond: A review. *Foods, 13*(18), 2948.

Xu, Y., Zhang, M., & Fu, L. (2023). Current application and limitations of augmented reality in the stomatology. *Chinese Journal of Stomatology, 58*(6), 592–597.

Zhang, S., & Yao, Z. (2025). The challenge of the application of augmented reality in science education in China: A systematic review. *Disciplinary and Interdisciplinary Science Education Research, 7*(1), 4.

CHAPTER 9

Future of Virtual and Augmented Reality

9.1 EMERGING TRENDS AND TECHNOLOGIES IN VIRTUAL AND AUGMENTED REALITY

As indicated in Table 9.1, unfolding technologies and trends are the way VR and AR continue to merge into other emerging categories, transforming government, education, entertainment, healthcare, and manufacturing in business. The direction of advancement is convergence, smartness, and sustainability – each trend by definition encompassing the creation of frictionless union of virtual and real worlds. The future of immersive technology will be as much about hardware innovation as it will be about designing a series of systems that can communicate with one another, interface with streams of real-time data, and collaborate in human-machine teams within large digital ecosystems.

Mixed Reality Convergence is the convergence of lines between virtual reality (VR) and augmented reality (AR) into blended spaces in which virtual and physical objects coexist and interact with one another in real-time. Convergence offers virtual and physical objects at the same time in order to facilitate more intuitive and natural interaction. MR finds its particular application in industrial training and maintenance through which workers are able to view within occluded machine components on the factory floor. In medical practice, MR is applied in pre-surgery planning and active medical training with the advantage of increased precision and reduced procedural risk. In product design and architecture, MR facilitates

DOI: 10.1201/9781003643395-9

TABLE 9.1 Emerging Trends and Technologies in Virtual and Augmented Reality

Emerging Trend/Technology	Description and Future Implications
Mixed Reality (MR) Integration	The integration of virtual and augmented reality in hybrid environments allows virtual and real objects to exchange information with each other without any restrictions. It is used today in factory training, medical simulation, and co-design group work systems.
AI-Driven Immersive Environments	Artificial Intelligence provides realism to VR/AR by using dynamic content generation, natural language interface, and predictive modeling of the user. Artificial Intelligence provides context-aware and customized virtual experiences.
Spatial Computing and Digital Twins	The combination of spatial computing and digital twin platforms provides real-time synchronization of virtual and physical objects. It is used in Industry 4.0 application, predictive maintenance, and visualization of smart cities.
Haptic and Multisensory Feedback Systems	Sensory immersion is also improved with the advancement of touch surfaces, temperature, and odour feedback, which leads to more effective training and improved user experience in simulation therapy and training.
Cloud and Edge-Based Rendering	Cloud VR/AR minimizes dependency on hardware by pushing computationally intensive workload onto a remote server. Edge computing achieves low-latency performance for mobile and shared scenarios.
Extended Reality (XR) Ecosystems	XR is a collective term encapsulating VR, AR, and MR in one technology. Next-gen innovations are centred on interoperability standards as well as cross-platform support for enterprise applications.
Metaverse and Persistent Virtual Worlds	Shared, distributed, and long-term construction of virtual worlds enables new social interaction, commerce, and learning paradigms. It also creates data ownership ethics and governance problems.
Neuro-Interface Technologies	Brain-computer interfaces (BCIs) enable direct neural control of virtual worlds. Next-generation applications can offer cognition rehabilitation, telepresence, and affective content adaptation.
Lightweight and Wearable Displays	Advancements in micro-LED and optical material technology result in low-latency, high-definition ergonomic headsets, and AR glasses. They will drive mass-market adoption.
Sustainable VR/AR Development	Next-generation VR/AR technologies give utmost priority to energy savings, recycling of waste, and ecodesign to achieve global sustainability goals and ESG compliance.

Source: Author's own work.

cross-disciplinary collaboration through the provision of opportunities for groups of people to work together as they walk through and comment upon 3D models in hybrid environments. This points toward the adoption of permanent hybrid physical–digital spaces that enable real-time collaboration and improve operational productivity (Matoug-Elwerfelli et al., 2025; Samala et al., 2025).

The Artificial Intelligence (AI) role in immersive environments is also prominent. VR/AR systems driven by AI contain machine learning ability with features of producing context-aware, adaptive content, anticipatory of the user's requirements, and interpreting voice or gesture through natural language processing. The intelligent layer transforms static environments into adaptive systems possessing learning and adaptation ability based on user input and constructs customized experiences. In education, AI makes intelligent adaptive virtual teachers possible; in business, it makes emotion-based product visualization possible; and in medicine, it makes the examination of real-time patient data during AR-aided surgery possible. The ultimate effect of the trend is the development of completely autonomous immersive environments with emotional intelligence, real-time learning, and cognitive interaction between human and artificial performers (Dormegny et al., 2025; Jiang et al., 2025).

Spatial Computing and Digital Twins is another sector of innovation. Both the technologies combined allow real-time synchronization of the virtual and physical models with day-and-night data sharing. This facilitates a multi-dimensional feedback loop of simulation, forecasting, and optimization of the physical systems. AR-photoed digital twins of industries and cities can simulate real-time operating conditions of machinery, transport networks, or power grid. This geospatially enabled monitoring capability can enable Industry 4.0 and Quality 4.0 philosophy-based decision-making as well as predictive maintenance. In smart cities, spatial computing presents the city's mobility as well as environmental information in immersive scenarios with more contextualized information for urban planners and empowering citizens for co-design (Jonek-Kowalska & Wolniak, 2023; Lampropoulos, 2025; Leite & Vieira, 2025; Jonek-Kowalska & Wolniak, 2025; Rosak-Szyrocka & Wolniak, 2025).

Immersiveness is also made possible by Multisensory Feedback Systems and Haptic. The more recent haptic interface technology, thermal perception, and smell feedback has stretched even further the spectrum of sensations to which one is subjected in VR. Not only does multisensory technology allow for a more real simulation sense, but it also improves learning and therapy. Muscle resistance can be regained in physical medicine by haptic gloves for neuromotor rehabilitation. In combat and flight simulation, in this case, haptic feedback adds realism and situational awareness. Secondly, adding sensory inputs such as smell or temperature deepens emotional engagement and memory, and such technologies form the basis of simulation-based training and psychotherapy. Sensor convergence is one of the directions towards complete embodied digital experience (El Choueiri et al., 2025; Samaden et al., 2025).

Growth in immersive technologies is also driven by Cloud and Edge-Based Rendering, which revolutionizes VR/AR content processing and delivery. Cloud VR/AR dispenses with local computing hardware needs with cloud servers doing the computation and rendering. Cloud VR/AR brings immersive experiences to all at low prices. Edge computing, on the other hand, shifts computation to the edge near the user for reducing latency and enabling real-time responsiveness as required in mobility or collaboration scenarios. The two-way approach ensures scalability, interoperability, and seamlessness of experience among devices. It enables new business use cases like cloud training, remote inspection, and virtual office work collaboration (Jiang et al., 2025; Toyin et al., 2025; Yanti et al., 2025).

Extended Reality (XR) Ecosystems is an integration of VR, AR, and MR under a single platform. XR ecosystems focus on interoperability, cross-platform compatibility, and mode switching between immersion modes sequentially. Their conceptualization streamlines business solutions, particularly in manufacturing, logistics, and learning, where there are several realities within one workflow. The consolidation is with the vision of interoperable XR platforms that will unite various devices and applications to talk and share data, facilitating scalability and inclusiveness. Upon maturity of the ecosystem, it will form a spatial integrated internet, consolidating all the physical and virtual assets into a smart immersive layer Dogru et al., 2025; Wilsa et al., 2025).

One of the manifestations of this more general change is the Metaverse and Persistent Virtual Worlds. They are shared virtual worlds that are decentralized, taking the idea of the internet to the level of completely three-dimensional spaces with their own social structures and economies. They enable new human interaction, business, and innovation, revolutionizing entertainment, education, and telework. They also bring new digital governance, ethics, data privacy, and virtual property rights challenges. Its success in the future will depend on the creation of open, secure, and fair frameworks for digital identity, user interaction, and resource allocation (Aviles-Castillo et al., 2025; Dubiel et al., 2025; Stecuła, 2023a; Stecuła, 2023b).

Another option is Neuro-Interface Technologies that provide humans with direct neural interfaces in an attempt to interact with digital systems. Brain-computer interfaces (BCIs) are the future of human-machine interaction in which humans interact with virtual worlds using their minds or feelings. Future potential applications include neurorehabilitation, cognitive therapy, and virtual telepresence. Likewise, emotion-based adaptation can also become reality in the near term using technology, where virtual worlds are generated and sculpted dynamically based on perception of emotion with profound impacts for psychology, learning, and human-computer symbiosis Fathi et al., 2025; Ghzaiel et al., 2025; Wei et al., 2025).

Large-scale adoption of immersive technology also awaits the creation of Lightweight and Wearable Displays. Breakthroughs in optical material, micro-LED, and holographic projection technology made the devices light, thin, and yet efficient but without sacrificing their genuineness as high resolution and quality field of view. Heavy AR glasses and mobile VR headsets make the immersive technology business, logistics, and individual communications application accessible for everyday use. Technologies must be prepared to go to the mass market, bridging the gap between test beds and widespread consumer adoption of technology (Sorita et al., 2025; Tao et al., 2025).

The future of immersive technology also lies in the need for Sustainable VR/AR Development. With eco-sensitivity a consideration in industrial design, the next generations of VR/AR systems will rely on low power use, recycling and reuse of equipment, and environmentally friendly and safe production. Green cloud, low-power display, and biodegradable material are green topics that align with international Environmental, Social, Governance (ESG) objectives. This planetary priority ensures ecologic responsibility with technologic advancement, and these interactive systems are ecologic sustainability leaders and stakeholders too.

New virtual and augmented reality technologies and trends are listed in Table 9.1.

9.2 INTEGRATION OF VIRTUAL AND AUGMENTED REALITY WITH OTHER EMERGING TECHNOLOGIES

VR and AR technologies have also developed extensively throughout the 21st century, and they have positioned themselves to be merged into other next-generation technologies grouped together to create the digital economy. Started as independent visualization technology, VR and AR have become part of the larger ecosystems like AI, IoT, big data analytics, blockchain, digital twins, and 5G/6G networks. The planned integration phase is likely to not only boost the capability of AR and VR systems but reshape paradigms of decision-making mechanisms, data interpretation, and user interaction in cyber-physical spaces.

AI integration with VR/AR provides the cognitive layer that enables interactive and personalized virtual experiences. AI is the cognitive capability that provides virtual experience with interactivity and personalization. With machine learning tools like deep neural networks, AR and VR worlds can monitor the user's actions, feelings, hand movements, and environmental inputs smoothly in an effort to serve adaptive real-time information. It therefore offers intelligent scene reconstruction, object recognition, and semantic comprehension in virtual realms. For example, artificial intelligence-based computer vision facilitates delivery of AR through real-world surface mapping and sensing and conversational interfaces to virtual worlds through natural language processing (NLP). Manufacturing through simulation of user training through user performance assessment and dynamic adjustment of levels of difficulty is made possible by AI. AR apps based on AI in medicine make it possible for surgeons while operating with predictive visualization of anatomical shape and thus improve accuracy and safety. Such synergy is also used in the development of autonomous virtual agents with human behaviour modelling and thus to the enhancement of didactic and therapeutic capabilities of immersion technologies (Gautam et al., 2025; Skotnicka-Zasadzień & Wolniak, 2026; Stecuła, 2024; Stecuła, 2025; Wolniak et al., 2024b; Xia et al., 2025; Yang et al., 2025).

Another area of convergence is the Internet of Things (IoT). Its intersection with IoT and VR/AR allows it to create intelligent and interactive digital twins, i.e., virtual copies of real systems, equipped in real time from sensors. Convergence is the technology behind the so-called 'spatial web' where virtual space and real space can't be distinguished. For example, in intelligent manufacturing, equipment operations are sensed by IoT sensors and overlaid in real time onto AR displays for predictive maintenance, anomaly detection, and process optimization. Field service technicians can use AR glasses and are capable of overlaying temperature, vibration, or power consumption metrics onto actual equipment such that they can respond to anomalies in real time. Similarly, in intelligent cities, IoT sensors monitoring traffic, air, or energy consumption can provide data into VR-based city dashboards that planners and policymakers can use to simulate various policy interventions in virtual simulations. Thus, when VR/AR is used along with IoT, it provides situational awareness, context-aware visualization, and data-driven decision-making, all three of which are the key drivers of Industry 4.0 and Quality 4.0 (Alves et al., 2025; Prahani et al., 2025).

The second ginormous digital transformation potential comes from the combination of VR and AR with 'Big Data Analytics'. Since a lot of interactive and sensor data is being created by immersive technologies, real-time analysis is essential. Big data architecture allows for a platform to make sense out of virtual experience mapping at scale to user behaviour trends, attention, or cognitive load. In commerce and advertising, VR environments can be queried to infer consumer purchasing behaviour and emotional response to products and assist in creating more effective campaigns. In learning, information from AR-based education systems could potentially enable instructors to track student performance and adapt content. Two-way trade between big data and VR/AR offers a feedback loop: data created by analytics is both a source and a sink for data on a continuous basis, learning from and improving it. This junction also presents new avenues to investigate computer ergonomics and human-computer interaction as all motion within the virtual world is now an event that must be recorded in the form of extra data (Devi et al., 2025; Rosa-Garcia et al., 2025; Stecuła & Naramski, 2025; Yang et al., 2025; Stecuła, 2024).

The blockchain process has also been significantly to be accredited for open and secure operation within the VR/AR environments. Blockchain's distributed nature provides integrity, authenticity, and traceability of data, which is necessary where assets and digital identity are involved. In virtual economies like the metaverse, blockchain is the norm of verifiable digital ownership of virtual objects in the form of non-fungible tokens (NFTs). Blockchain makes secure identity management easy to safeguard so that users can have enduring interoperable digital identities on various VR/AR platforms. For commerce, blockchain provides guaranteed sensor data for AR maintenance or

VR simulation which is tamper-proof and hence there is enhanced compliancy and auditability. Smart contracts also provide transaction automation in VR markets or service platforms and minimize the intermediary role. The intermarriage of blockchain and immersive technology therefore turns the virtual world into an institutionalized, trusted, secure environment where digital exchange and interaction of value become institutionalized (Ball et al., 2025; Fathi et al., 2025; Zhu & Su, 2025).

The introduction of 5G and the forthcoming 6G networks introduced communications infrastructure supporting real-time high-fidelity VR and AR experiences. They also offer ultra-low latency, high capacity, and huge device connectivity – settings where massive-scale IoT integration and shared immersive experience is the whole ask. Real-time synchronous shared experience is enabled by 5G in virtual offices where multiple consumers are able to react to virtual objects in real time and not even realize there is any delay. This is second nature to telework, telemedicine, and distant learning, where presence and immediacy are the requirement. In the not-so-distant future, 6G networks with edge computing and AI-native network architecture will further upscale VR/AR quality with distributed intelligence and context-aware service deployment. Render jobs, for example, can pre-load onto edge servers and offload a part of computation workload of local devices without affecting responsiveness. This is the way infrastructure convergence indicates there is no development of immersive technologies without there being communication networks in place (Fathi et al., 2025; Wilsa et al., 2025).

Yet another one of the most promising fields where convergence of technologies is going to occur in the near future is convergence of digital twin with VR/AR. Virtual replicas of physical objects as end-to-end streams of information updates become interpretation and visualization capability using immersive interfaces. Engineers even communicate with virtual models of factories using VR headsets, observing interdependencies of complex systems and experimenting with the effects of change in operations within the realm of a risk-free environment. Digitally twinning patients whose vital signs are tracked by means of AR in medicine enables doctors to model operations or forecast the progression of disease. This convergence strengthens not only information of sophisticated systems but incremental decision-making and resilience of foresight too. Digital twin and VR/AR convergence is the evolution of intelligent systems from experience simulation (dynamic) to static modelling (Jiang et al., 2025;

Jiang et al., 2025; Samala et al., 2025; Wolniak & Stecuła, 2024; Wolniak et al., 2024a).

Convergence with edge AI, cloud computing, and robots is also predicted to achieve the highest potential of immersive technology. AR interfaces enable natural robot operation and visual monitoring of robot activity, and VR environments enable remote monitoring by operators of autonomous executions conducted under undesirable conditions. Cloud and edge computing infrastructure naturally facilitates scalable VR/AR solutions with minimal effort by tapping into the rendering, simulation, and analytics compute power in distributed environments. Edge AI is particularly placing intelligence near where data are being generated, minimizing latency, and providing privacy-protected analytics – a highly desired application in defense and healthcare application spaces.

9.3 ETHICAL AND SOCIETAL IMPLICATIONS OF FUTURE VIRTUAL AND AUGMENTED REALITY ADOPTION

Immersive technology expansion has an unimaginably enormous number of ethical, psychological, and sociocultural issues many times greater than the potential of technology (Table 9.2). These are core issues, not marginal issues to ethical digital ecosystem design in the sense that they are issues dealing with human dignity, cognitive health, and common social values. The convergence of virtual technologies and artificial intelligence, biometric information, and networked space is a setting in which ethical governance ought to be attempting to keep up as quickly as technological potential itself. All these subject areas handled here presume how far VR and AR will engage significant aspects of the human experience – ranging from agency and identity through law, sustainability, and culture.

Privacy and data protection are most critical among these. VR and AR technologies are based on real-time capture and processing of highly personal data in an attempt to create responsive and realistic environments. Data include spatial mapping, biometric feedback like eye movement, heartbeat, facial expressions, and behaviour patterns that reveal cognitive and emotional states. Unlike conventional digital information, these experience databases expose individuals' innermost sides, making it practically feasible for platforms to introspect on a user's cognitive and physical map. Absent draconian limitations on data protection, real threats of exploitation, manipulation, or wiretapping hang in the air. Future VR/AR creation will therefore require holistic regulatory frameworks, encryption policy, and consent policy on the same level as for

TABLE 9.2 Ethical and Societal Implications of Future Virtual and Augmented Reality Adoption

Area of Concern	Description	Future Implications and Considerations
Privacy and Data Security	Virtual/augmented reality worlds constantly collect biometric, spatial, and behaviour data such as gaze, movement, and emotional feedback. Data such as these are entirely central to adaptive immersion but extremely sensitive in nature.	Future legislation needs to incorporate strong data protection, encryption, and consent mechanisms to prevent state and commercial exploitation abuse and espionage.
Digital Identity and Representation	Users can tailor avatars and virtual characters that can be fairly divergent from individuals they just happen to be in the real world, leading to issues regarding authenticity, accountability, and deception.	Virtual identity management processes need to be incorporated into metaverse and extended reality governance structures that prevent theft, impersonation, and discriminatory impersonation within the metaverse and extended reality.
Psychological and Cognitive Effects	Mass usage of immersive environments can shape individuals' constructions of reality, emotional equilibriums, and social conduct. Addictions, desensitizations, and dissociations were reported in initial experiments.	Long-term psychological research and policy will need to be developed in a way that monitors cognitive wellness and creates responsible content production that maintains psychological integrity.
Social Inequality and Accessibility	Immersive technologies are hardware-dependent and costly, which may sustain or widen digital divides shaped by socioeconomic status and geographic location.	Low-cost hardware, universal access, and accessible design approaches will need to be enabled by policy and design in the future to prevent new exclusionary technologies.
Ethical Design and Manipulation of Reality	The increasing sophistication of VR and AR in replicating real-world environments and eliciting emotions raises ethical concerns related to manipulation, persuasion, and disinformation.	Labeling of content for transparency, responsible algorithms, and responsible standards for immersive advertising and learning will need to be established in order to maintain autonomy and trust.

(Continued)

TABLE 9.2 Ethical and Societal Implications of Future Virtual and Augmented Reality Adoption *(Continued)*

Area of Concern	Description	Future Implications and Considerations
Workplace Transformation and Human Agency	AR-augmented workplaces and virtual training change workplaces, potentially expanding surveillance and reducing human agency.	Employee dignity, privacy, and co-determination will have to be ethically founded for algorithmic decision-making in industry-immersive workplaces.
Cultural and Social Identity	Immersive space facilitates cultural expression and co-creation but carries risks of cultural appropriation, prejudice, and experience homogenization.	Diversity and cultural protection policies will need to make virtual worlds pluralist, tradition-respecting, and authentic.
Legal Accountability and Governance	The virtual/real-world action dichotomy is eroded incrementally, rendering current legal definitions of harm, ownership, and responsibility increasingly dubious.	International institutions and government will need to reconsider paradigms of digital law with a view to including VR/AR-concerned behaviour, liability, and intellectual property rights.
Environmental Sustainability	Energy usage related to immersion to facilitate immersive technology, server rendering, and data centres contributes to the environmental load.	Green computing values, energy efficient design, and circular economy practices with ESG guidelines need to be considered in the future development.
Human-Technology Dependency	Greater immersion with immersion technology in daily life can enhance dependence and reduce social interaction and sympathy in the real world.	Ethics and education should foster balanced use, information literacy, and psychological resilience in order to preserve authentic and sustainable human relationships in a hybrid-reality world.

Source: Author's own work.

medical or financial data. Ethical design must provide that never shall personalization intent be greater than the user control right and privacy (Chauhan & Kumar, 2025; Goldust & Grant-Kels, 2025; Holohan et al., 2025; Lui et al., 2025).

There is yet another ethical concern in virtual representation and identity concerns. Immersive space permits avatars or virtual identities quite dissociated from bodies. As flexibility encourages accessibility and creativity, it also leads to deception, impersonation, and psychological dissonance. Distinguishing genuine from simulated actors is more complicated in virtual multi-user environments. Accountability is challenged where verification and trust are issues of utmost concern in professional, pedagogic, or business transactions. Regulator interest will thus be compelled to balance regimes of identity control offering no anonymity but no unaccountability as well. Conventional designs for digital personhood are now finally feasible to construct – e.g., passports or certificates – to secure moral continuity of virtual contact between locations (Hung et al., 2025; Karimov et al., 2025).

No less relevant are cognitive and psychological impacts of prolonged immersion in simulated worlds. Studies already indicate that prolonged exposure to simulated worlds can affect emotion regulation, attention, and social perception. VR experiences that can evoke powerful affective responses are powerful, in the hands of the wrong person, for desensitizing or addicting individuals. The same level of immersion that makes VR such a useful tool for instruction or therapy is the same level of immersion that will be abused in worst-case scenarios of psychological manipulation. Ethical creators should thus collaborate multidisciplinary with psychologists and neuroscientists in creating mental health protection, exposure guidelines, and cognitive well-being-focused content guidelines. Therapeutic immersion instead of behaviour conditioning in the coming decade will squarely depend on such ethical boundaries (Koskinen et al., 2025; Raja & Al-Baghli, 2025).

Diffusion of VR/AR also broaches social inequality and access issues. Unaffordable costs of equipment, bandwidth intensity, and technological capability barriers threaten to keep immersive technologies within the reach of elite fractions alone. It would be able to sustain health inequalities, work inequalities, and educational inequalities in a digital divide of this nature. It is the moral obligation that drives technology designers and policymakers to adopt inclusive design approaches to include such technology into economic accessibility and affordability within diverse groups. Universal design requirements, open platforms, and public acceptance of digital inclusion initiatives will be required for fair opportunity. Without them, immersive systems can reproduce, not challenge, social injustices (Asif et al., 2025; Kwok et al., 2025).

The ability of VR and AR to make us perceive things as other than they are introduces a new ethical challenge: the manipulation of reality itself. With the disappearance of boundaries between worlds real and virtual, artists get a new degree of control over users' emotion, cognition, and choice. Political engineering, interactive promotion, or education can shape intellectual processes more than our media do, perhaps by avoiding critical thinking. Translucent labelling of virtual content, algorithmic transparency, and open ethics for immersion media are required to provide user agency. Accurate representation and avoidance of simulated immersion should be integrated into virtual design professional ethics (Musasa, 2025; Palakurti, 2025).

VR and AR are reshaping professional space, inducing multidimensional work transformations. More sophisticated industrial or corporate frameworks can themselves enhance productivity and security but enable close tracking of work performance by data. When wearable technology, for example, is applied in gesture tracking, eye fixation, or physiological signals on a full-time basis, the boundary is crossed between helping and surveillance. Ethical use would therefore need to maintain workers' autonomy, privacy, and dignity. Openness guidelines for algorithmic testing procedures and the co-determination principle need to rule the application of immersion technologies to human-machine cooperation systems. Ethically, responsible decision-making will determine if the use of such technology promotes or retards autonomy (Asif et al., 2025; Goldust & Grant-Kels, 2025; Karimov et al., 2025).

Cultural adoption issues for VR and AR are also paramount. Immersive experiences permit unimagined amounts of creativity and intercultural exchange while presenting challenges to cultural identity and appropriation. Medias distributed globally in virtual space hold the power to homogenize experience of culture and substitute local culture. In addition, algorithmic bias in recommending content or appearance of avatars can reinforce stereotypes. Livable immersive culture will therefore comprise interest in diversity, authenticity, and context sensitivity. Producers and institutions have a duty to develop virtual worlds that are pluralism-sensitive and seek intercultural dialogue rather than cultural hegemony (Kour, 2024; Perera et al., 2024).

Legal control and responsibility is another important aspect of ethical concern. Cyberbullying, theft of virtual items, or propagating abusive content in the virtual world makes the traditional physical presence-based legal system even heavier. The muddled distinction between simulated and real harm inhibits law enforcement in its current form. Virtual ownership of property, authorship of machine-generated content, and rights on data about one's self in immersive worlds must be established by law for the first time. Governance regimes will be required to follow up, establishing norms for accountability, intellectual property, and virtual citizenship in hybrid reality spaces. Developing standardized legal constructs will be the foundation upon which justice and order will be provided to future metaverse economies (Pawar & Pawar, 2024; Ursin et al., 2024).

Environmental sustainability is a below-the-radar one that is no less critical. Rendering, data transportation, and production processes are energy-hungry and deeply environmentally expensive. Taking VR and AR as two distinct entities, both of them can certainly accelerate environmental degradation. Green energy utilization, product recycling, and sustainable supply chain are the cornerstones of sustainable VR/AR theory of development. Sustainable system design and operation with ESG considerations ensure world well-being and development go hand in hand. With a greener digital economy beckoning on the horizon, its contribution to sustainability will be a measure of performance along with user experience or computing power (Ezzeldin et al., 2024; Wieczorek & Stecuła, 2025).

Greater immersion of immersive technologies carries the risk of human-technology reliance. As VR and AR seize communication, learning, and entertainment, people start replacing face-to-face interaction with virtual interaction. Such reliance on virtual worlds will drain social solidarity, sympathy, and common identity. Dwelling in them will take effortful self-regulation, technological literacy, and technology-practice based on restraint. Schools must become adept at instructing people to switch between worlds without psychic discombobulation. Man's future in the sense of successful life in an immersed age will rely on achieving a balance between technology-augmentation and genuine social presence.

BIBLIOGRAPHY

Alves, C., Machado, J., & Reis, J. L. (2025). Impacts of augmented reality in e-commerce: Advances and trends in 2024—A literature review. *Smart Innovation, Systems and Technologies*, *438*, 931–943.

Asif, M., Fazel, H., & Alghamdi, F. S. (2025). The ethical landscape of immersive marketing: Virtual and augmented reality in advertising. *Marketing Intelligence and Planning.* https://www.emerald.com/mip/article-abstract/43/7/1515/1255003/The-ethical-landscape-of-immersive-marketing?redirectedFrom=fulltext

Aviles-Castillo, F., Buele, J., & Palacios-Navarro, G. (2025). Virtual reality and user experience: Current trends and future challenges. *IEEE Access*, *13*, 55939–55956.

Ball, C., Huang, K.-T., & Francis-Levin, J. (2025). Beyond the hype: Longitudinal trends in virtual reality perceptions beyond the COVID-19 pandemic. *Information Research*, *30*(iConf(2025)), 260–279.

Chauhan, S., & Kumar, V. (2025). Virtual horizons: Navigating legal and ethical frontiers in HR virtual reality technology. In: *Future of HRM in a world of persistent virtual reality* (pp. 167–195). IGI Global.

Devi, R. S., Kavitha, K., & Usha, D. (2025). Contemporary trends and usage of augmented reality tools for autistic children. *Lecture Notes in Networks and Systems*, *1399*, 99–110. https://link.springer.com/chapter/10.1007/978-3-031-91005-0_10

Dogru, M. S., Yüzbasioglu, F., & Arpaci, I. (2025). A bibliometric trend analysis of augmented reality in STEM education. *Innovations in Education and Teaching International.*

Dormegny, L., Yaïci, R., Koestel, E., Lejay, A., & Bourcier, T. (2025). Global trends and practice patterns in virtual reality simulation training for ophthalmic surgery: An international survey on the use of virtual reality simulation training around the world. *Scientific Reports*, *15*(1), 30886.

Dubiel, A., Kamińska, D., Zwoliński, G., Agostini, D., & Zancanaro, M. (2025). Virtual reality for the training of soft skills for professional education: Trends and opportunities. *Interactive Learning Environments*, *33*(5), 3261–3281.

Elaish, M. M., Yadegaridehkordi, E., & Ho, Y.-S. (2025). Publication performance and trends in virtual reality research in education fields: A bibliometric analysis. *Multimedia Tools and Applications*, *84*(10), 6999–7026.

El Choueiri, J., Pellicanò, F., Di Cosmo, L., Elmi-Terander, A., & Cannizzaro, D. (2025). Augmented reality in spinal neurosurgery: A bibliometric analysis of trends and clinical applications. *World Neurosurgery*, *200*, 124204.

Ezzeldin, A., Ayman, N., Tawfik, O., Ragheb, F., & Makramalla, M. (2024). Socio-ethical challenges of integrating augmented reality into transdisciplinary engineering programs. *Advances in Transdisciplinary Engineering*, *60*, 749–758.

Fathi, S., Sabeti, S., Shoghli, O., Heydarian, A., & Balali, V. (2025). Adoption of virtual and augmented reality in the architecture, engineering, construction, and facilities management (AEC-FM): Mixed method analysis of trends, gaps, and solutions. *Frontiers in Built Environment*, *11*, 1580639.

Gautam, A., Asutkar, S., & Nagpure, D. (2025). Virtual reality in surgical training: Trends and prospects. *Multidisciplinary Reviews*, *8*(5), e2025136.

Ghzaiel, K., Bouzaabia, R., & Hassairi, M. (2025). When the artificial revolutionizes the reality: Focus on this new trend of virtual influencers. *Lecture Notes in Business Information Processing*, *531*, 309–328.

Goldust, M., & Grant-Kels, J. M. (2025). Regulatory considerations for safe and ethical use of augmented reality and virtual reality in dermatology. *Clinics in Dermatology*, *43*(2), 284–286.

Holohan, C., Hogan, M., Kelly, D., Kennedy, M., & Silke, C. (2025). Ethical witnessing: Participatory virtual reality production and the experience of homelessness. *Studies in Documentary Film*. https://www.tandfonline.com/doi/full/10.1080/17503280.2025.2470117

Hung, L., Suchomelová, V., Diallo, K., Wong, J. O. Y., & Ren, L. H. (2025). Ethical considerations in implementing virtual reality programs in long-term care settings: Case studies from Canada and the Czech Republic. *Ethics, Medicine and Public Health*, *33*, 101074.

Jiang, H., Zhu, D., Chugh, R., Turnbull, D., & Jin, W. (2025). Virtual reality and augmented reality-supported k-12 STEM learning: Trends, advantages, and challenges. *Education and Information Technologies*, *30*(9), 12827–12863.

Jonek-Kowalska, I., & Wolniak, R. (2023). *Smart cities in Poland: Towards sustainability and a better quality of life?* Routledge.

Jonek-Kowalska, I., & Wolniak, R. (2025). *Exclusion in smart cities: Assessment and strategy for strengthening resilience*. Routledge.

Karimov, A., Ismayilzada, A., & Saarela, M. (2025). Ethical and legal implications to data collection in virtual and augmented reality. *2025 IEEE Symposium on Trustworthy Explainable and Responsible Computational Intelligence (CITREX Companion 2025).*

Koskinen, C., Dovland Vassbø, A. G., & Estman, L. (2025). A caring-perception model for ethical competence in virtual reality environment. *Nursing Ethics*. https://pubmed.ncbi.nlm.nih.gov/40103534/

Kour, M. (2024). Ethical considerations in the metaverse and augmented retail reality era. In: *Consumer experience and decision making in the metaverse* (pp. 99–111). IGI Global.

Kwok, A. P. K., Mui, S. Y., & Yan, M. (2025). Enhancing university students' awareness of AI ethical issues using virtual reality: A preliminary study. *Smart Innovation, Systems and Technologies*, *432*, 383–394.

Lampropoulos, G. (2025). Combining artificial intelligence with augmented reality and virtual reality in education: Current trends and future perspectives. *Multimodal Technologies and Interaction*, *9*(2), 11.

Leite, H., & Vieira, L. R. (2025). The use of virtual reality in human training: Trends and a research agenda. *Virtual Reality*, *29*(1), 25.

Lui, P. P., Stringer, E., & Jouriles, E. N. (2025). Using virtual reality in discrimination research: Ethical considerations and participants' perspectives. *Journal of Empirical Research on Human Research Ethics*, *20*(4), 179–189.

Matoug-Elwerfelli, M., Daud, A., Ali, K., & Abdou, A. (2025). Virtual reality in restorative dentistry: A bibliometric analysis of research trends. *BDJ Open*, *11*(1), 77.

Musasa, T. (2025). Ethical implications of artificial intelligence (AI) and augmented reality (AR) in marketing research. In: *Revolutionizing academic research with AI and augmented reality* (pp. 397–410). IGI Global.

Palakurti, A. (2025). Harnessing AI and augmented reality for academic research: Innovations, applications, and ethical challenges. In: *Revolutionizing academic research with AI and augmented reality* (pp. 33–66). IGI Global.

Pawar, V. V., & Pawar, D. (2024). Augmented reality in medical education: Exploring ethical considerations. *Oral Oncology Reports*, *9*, 100208.

Perera, K., Regenbrecht, H., Langlotz, T., & Pantidi, N. (2024). Ethical implications of pervasive augmented reality. *MobileHCI 2024 Adjunct Proceedings: Publication of the 26th International Conference on Mobile Human-Computer Interaction*, 37.

Prahani, B. K., Dawana, I. R., & Kurtuluş, M. A. (2025). Trend analysis of augmented reality in science education for supporting SDGs. *Perspektivy Nauki i Obrazovania, (3)*, 676–691. https://pnojournal.wordpress.com/wp-content/uploads/2025/07/pdf_250345.pdf

Raja, U. S., & Al-Baghli, R. (2025). Ethical concerns in contemporary virtual reality and frameworks for pursuing responsible use. *Frontiers in Virtual Reality, 6*, 1451273.

Rosa-Garcia, A. D. L., Marrufo, ÁI. S., Luviano-Cruz, D., Rodriguez-Ramirez, A., & Garcia-Luna, F. (2025). Bridging remote operations and augmented reality: An analysis of current trends. *IEEE Access, 13*, 36502–36526.

Rosak-Szyrocka, J., & Wolniak, R. (2025). *Smart universities in smart cities: Shaping the future of education and urban innovation*. Routledge.

Samaden, I. S., Ahmad, I., & Salam, S. (2025). Technology trends and augmented reality: A systematics literature review. *Journal of Advanced Research in Applied Sciences and Engineering Technology, 53*(1), 129–140.

Samala, A. D., Rawas, S., Rahmadika, S., Fikri, R., & Sandra, R. P. (2025). Virtual reality in education: Global trends, challenges, and impacts—Game changer or passing trend? *Discover Education, 4*(1), 229.

Skotnicka-Zasadzień, B., & Wolniak, R. (2026). Artificial intelligence in public administration. In: *Routledge studies in innovation, organizations and technology*. Routledge. https://www.routledge.com/Artificial-Intelligence-in-Public-Administration/Skotnicka-Zasadzien-Wolniak/p/book/9781032879284

Sorita, E., Duclos, N. C., Puech, L., Chibaudel, Q., & Quillion-Dupré, L. (2025). Virtual reality in clinical evaluation of unilateral spatial neglect, anatomy of current trends: A scoping review. *Applied Neuropsychology: Adult*. https://pubmed.ncbi.nlm.nih.gov/39780726/

Stecuła, K. (2023a). Identifying diverse uses of virtual reality in higher education and exploring perceptions of VR in the chosen field. *Zeszyty Naukowe Politechniki Śląskiej. Organizacja i Zarządzanie*. https://managementpapers.polsl.pl/wp-content/uploads/2023/12/181-Stecu%C5%82a.pdf

Stecuła, K. (2023b). Review of virtual reality applications applied in industrial practice. *Zeszyty Naukowe Politechniki Śląskiej. Organizacja i Zarządzanie*. https://managementpapers.polsl.pl/wp-content/uploads/2024/01/187-Stecu%C5%82a.pdf

Stecuła, K. (2024). Analysis of asymmetric VR games—Steam platform case study. *Technology in Society, 78*, 102673.

Stecuła, K. (2025). Advancing education through virtual reality in the management and production engineering field of study. *Zeszyty Naukowe Politechniki Śląskiej. Organizacja i Zarządzanie*. https://managementpapers.polsl.pl/wp-content/uploads/2025/05/215-Stecu%C5%82a.pdf

Stecuła, K., & Naramski, M. (2025). Virtual reality as a green tourism alternative: Social acceptance and perception. *Sustainability, 17*(17), 7722.

Stecuła, K., Wolniak, R., & Aydın, B. (2024). Technology development in online grocery shopping—From shopping services to virtual reality, metaverse, and smart devices: A review. *Foods, 13*(23), 3959

Tao, L., Cukurova, M., & Song, Y. (2025). A systematic review of learning analytics in immersive virtual reality: Trends, challenges, and implications. *CEUR Workshop Proceedings, 3995*, 41–48.

Toyin, J. O., Sattineni, A., Wetzel, E. M., Fasoyinu, A. A., & Kim, J. (2025). Augmented reality in U.S. construction: Trends and future directions. *Automation in Construction, 170*, 105895.

Ursin, F., Timmermann, C., Benzinger, L., Salloch, S., & Tietze, F.-A. (2024). Intraoperative application of mixed and augmented reality for digital surgery: A systematic review of ethical issues. *Frontiers in Surgery, 11*, 1287218.

Wei, Y., Chen, Y., Ma, R., Zhang, L., & Gao, Q. (2025). Research hotspots and trends of virtual reality intervention for stroke: Bibliometric analysis. *JMIR Serious Games, 9*, e6599.

Wieczorek, A., & Stecuła, K. (2025). Supporting sustainable demographic change: Augmented reality in men's sheds as a concept for active aging. *Sustainability, 17*(18), 8345.

Wilsa, A. W., Hartono, Subali, B., & Rahayu, E. S. (2025). Research trends of augmented reality in biology learning: Content and bibliometric mapping analysis. *Malaysian Journal of Learning and Instruction, 22*(1), 157–180.

Wolniak, R., Gajdzik, B., Grebski, M., Danel, R., & Grebski, W. (2024a). Business models used in smart cities—Theoretical approach with examples of smart cities. *Smart Cities, 7*(4), 1626–1669.

Wolniak, R., & Stecuła, K. (2024). Artificial intelligence in smart cities—Applications, barriers, and future directions: A review. *Smart Cities, 7*(3), 1346–1389.

Wolniak, R., Stecuła, K., & Aydın, B. (2024b). Digital transformation of grocery in-store shopping—Scanners, artificial intelligence, augmented reality and beyond: A review. *Foods, 13*(18), 2948.

Xia, Y., Pang, P. C.-I., Xiong, Z., Liu, X., & Liu, T. (2025). Virtual reality technology in digital health education: Current research status and future trends. *Communications in Computer and Information Science, 2447*, 53–66.

Yang, Y., Rasul, M. S., & Jie, L. (2025). Bibliometric analysis of trends and applications of augmented reality and virtual reality in vocational education. *Journal of Computational Methods in Sciences and Engineering.* https://journals.sagepub.com/doi/10.1177/14727978251338471

Yanti, F., Lufri, L., & Ahda, Y. (2025). Current trends in augmented reality to improve senior high school students' skills in education 4.0: A systematic literature review. *Open Education Studies, 7*(1), 20240053.

Zhu, Y., & Su, Y.-S. (2025). Virtual reality trends and challenges in teacher education: A systematic literature review. *Interactive Learning Environments.* https://www.tandfonline.com/doi/full/10.1080/10494820.2025.2482585

Index

3D, 2, 5–6, 20–22, 25, 33, 35–38, 41–43, 45–47, 55–56, 67, 70–71, 73–75, 85, 102–104, 139

A

Accessibility, 66–67, 83, 93, 104, 108, 119, 121, 128, 130, 146, 148
Addiction, 30–31, 68–69, 74, 121–122, 127–128, 130
Application, 6–12, 22, 24–25, 28, 33, 35, 39–41, 43–44, 46–47, 51–52, 55, 57, 63–65, 70–72, 81, 83, 85, 104, 106–107, 119, 122, 126–127, 131, 137–138, 141, 145, 149
AR application, 47, 55, 57, 72, 85, 127
AR goggles, 6–8, 23–25, 58, 74
Artificial intelligence, 41, 88, 126, 138–139, 142, 145
Augmented reality, 1, 5, 9–10, 12, 23, 28, 30, 34, 40, 45, 55, 59–60, 74, 80, 83, 89–90, 97, 100–103, 106–107, 111, 118, 120–122, 124, 137–138, 141, 145–147

C

Controllers, 1, 7, 18, 20–21, 25, 33, 35, 53, 63
Culture, 11–12, 18, 21–22, 27, 30, 35, 60, 66, 68, 75, 82, 97–101, 108, 110–111, 119, 124, 145, 149

E

Education, 1, 6, 10–11, 25, 30, 60, 66–67, 74, 83, 99, 101–104, 109–111, 119, 127, 130, 137, 139, 143, 147, 149
Emotions, 1, 27, 29–31, 42, 45, 53, 85, 146
Engagement, 5, 22, 27–29, 31, 33–37, 43–44, 46, 56, 70, 73, 98, 102, 110, 122, 127, 140
Entertainment, 1, 11–12, 18, 21–23, 25, 27–30, 57, 80, 97, 100–101, 104, 110, 119, 121, 137
Ethics, 88, 124, 138, 141, 147, 149
Experience, 1–3, 5, 20–23, 25–31, 34–37, 43–46, 55–56, 59, 63, 65–66, 72–73, 81–83, 87–88, 91–92, 97–100, 102, 104–110, 118, 121–122, 128–130, 138, 140, 142–145, 147, 149

G

Games, 8, 18–23, 25, 27–28, 54, 60, 69, 71, 80, 104

H

Haptic gloves, 21, 53, 66, 140

I

Immersion, 3, 5, 7, 20, 22, 26, 28–31, 34, 45, 53, 69–70, 82, 84–85, 90–92, 104–105, 107, 118, 120–121, 124–126, 128, 130, 138, 140, 142, 146–149

Immersive technologies, 1, 3, 28, 30, 52, 59–60, 63, 97, 106, 109–110, 125–129, 140, 143–144, 146, 148
Industry, 6, 8–10, 22, 34, 36–38, 42, 45–46, 59, 66–69, 104–105, 129–130, 138–139, 143
Innovation, 59, 68, 80, 94, 101, 103–104, 106, 108, 110, 119, 122, 126, 128, 130, 137, 139–140
Interface, 137–139, 141
Isolation, 31, 54, 68–69, 103, 120–123, 127, 130

M

Marketing, 10, 45–46, 98, 105, 110
Mental health, 68, 81, 83, 87–88, 92, 148
Mixed reality, 4, 97, 108, 137–138
Museums, 6, 11, 18, 21, 27, 63–64, 66–67, 71, 74, 97, 101, 104, 108, 111

P

Physical activity, 23–25, 51, 53–54, 59, 70
Privacy, 22, 27, 74, 83, 88, 100, 122–123, 125, 141, 145–147, 149

R

Rehabilitation, 10, 100, 103, 106, 109, 111, 138, 140
Relaxation, 68, 80–92, 97, 99–101, 103
Retail, 27, 33–40, 44–45, 97–98, 100–102, 105
Risks, 27, 30, 54, 69, 87, 120–122, 124–125, 127–128, 130, 147

S

Safety, 9, 11, 22, 24, 28, 59, 86, 108, 111, 118, 142
Sensors, 3, 6–7, 21, 24, 41, 53, 81, 83, 92, 119, 121, 142–143
Shopping, 6, 11, 25–26, 33–41, 43–46, 78, 97–98, 100–102, 105, 111
Simulation, 18–19, 46, 51, 54, 67, 101–103, 105, 109–111, 138–140, 142, 144–145
Smart city, 59, 107–110
Social interaction, 92, 100–101, 103–104, 106, 123, 138, 147
Sport, 11, 21, 25, 51–52, 55, 60, 97, 99–102

T

Tourism, 10–11, 22, 63, 65–71, 73–74, 97, 99, 101–102, 105, 110
Training, 8–10, 19, 51, 53–55, 57, 59, 68–69, 81, 83–85, 89, 98–106, 109, 111, 125, 129, 137–138, 140, 142, 147

U

User experience, 31, 36, 46, 138

V

Virtual environment, 3–4, 18, 36, 53, 82–83, 99, 111
Virtual reality, 1–5, 21, 33, 38, 51, 60, 63, 65, 80, 85, 90, 97, 101–103, 111, 118, 137
Virtual tourism, 22, 63, 66, 70
Virtual training, 53–54, 59, 147
Virtual world, 2–3, 5, 8, 21, 28–31, 74, 80, 84, 104, 120, 122, 124, 143–144, 149
VR application, 51–52, 63–65, 85
VR goggles, 2, 8, 20, 35, 51–52, 63, 65

For Product Safety Concerns and Information please contact our EU
representative GPSR@taylorandfrancis.com
Taylor & Francis Verlag GmbH, Kaufingerstraße 24, 80331 München, Germany

www.ingramcontent.com/pod-product-compliance
Ingram Content Group UK Ltd.
Pitfield, Milton Keynes, MK11 3LW, UK
UKHW020206200926
475702UK00004B/643
9781041080282